MW00849144

The Biological Basis of Teleological Concepts

Harry Binswanger

The Ayn Rand Institute Press

Los Angeles, California

Copies of this book may be purchased from the publisher.
All inquiries should be addressed to: The ARI Press,
4640 Admiralty Way, Suite 715, Marina del Rey, CA 90292

ISBN: 0-9625336-0-2

Library of Congress Catalog Card Number: 90-080162

Manufactured in the United States of America

Editorial/production supervision by Donna Montrezza

CONTENTS

CONTENTS

CONTENTS

PREFACE

The present book is an edited version of my doctoral dissertation, which was written from 1969 through 1973. In the ensuing years, the subject of teleology has received much attention in the literature, and by way of updating the dissertation I have added an appendix assessing the most important of these newer works. The body of the book presents the same argument in the same progression as in the dissertation.

A word of explanation is required concerning the later chapters, where I deal with objections to my position. There are such things as unreasonable objections—objections stemming from a wrong approach, invalid presuppositions, context-dropping, etc.—and these are properly ruled out of philosophical court rather than rebutted. I have, however, retained several sections rebutting objections of this type on the grounds that in rebutting them I make some points of independent value. But I want to be on record that I do not believe that raising or answering "counterexamples"—the bread and butter of analytic philosophy—is a necessary or productive enterprise, assuming that sufficient direct evidence for one's position has been provided.

In the present case, that direct evidence is as solid and as incontrovertible as the distinction between life and death. I can say this in all modesty, because the distinction between life and death is the very one on which my theory is based. For my thinking on teleology was inspired by Ayn Rand's validation of man's life as the standard of morality, and that validation hinges on her analysis of the epistemological roots of the concept of "value":

> It is only the concept of "life" that makes the concept of "value" possible. . . . An *ultimate* value is that final goal or end to which all the lesser goals are the means. . . . It is

i

only an ultimate goal, an *end in itself*, that makes the existence of values possible. Metaphysically, *life* is the only
phenomenon that is an end in itself: a value gained and
kept by a constant process of action. Epistemologically,
the concept of "value" is genetically dependent upon and
derived from the antecedent concept of life. To speak of
"value" as apart from "life" is worse than a contradiction
in terms ("The Objectivist Ethics," *The Virtue of Selfishness*).

My philosophical debt to Ayn Rand goes far beyond her
theory of value (only a highly excerpted portion of which is
given above). Her Objectivist philosophy is the framework for
everything I have written or will write. There were several references to her writings in my original dissertation, but I am
happy to be able to add more for the present edition.

INTRODUCTION

This work is a sustained argument in favor of an exclusively biological teleology. There is an inescapable connection, I will argue, between life and goals. The actions of living organisms, even of non-conscious organisms such as plants, are goal-directed; the actions of inanimate objects, even of man-made devices such as "target-seeking" torpedoes, are not.

In defending this thesis, my approach is epistemological, not linguistic: I am seeking to identify the *proper* scope of teleological concepts. Conventionalists may here complain that decisions about the proper scope of concepts have to be made by us, since "nature does not tell us" how to classify phenomena. But the defeat of Platonism does not establish conventionalism: it is still up to us to divide nature at its joints—i.e., according to the fundamental similarities and differences in the subject matter. (The conventionalist objection is rebutted at length in chapter 10.)

In deciding the proper scope of teleological concepts, the first step is to ask: what phenomena in nature gave rise to the formation of teleological concepts? What distinctions were these concepts originally intended to mark? This will involve focusing on some paradigm cases of teleology—simple, clear, fairly non-controversial instances of action for the sake of a goal.

Then I consider what becomes of these concepts in the light of a more sophisticated context of knowledge. I address the question of whether or not teleological concepts mark any essential distinction in nature, and, if so, precisely what distinction. Are the facts on the basis of which teleological concepts were first formed reducible to other facts of a more fundamental order? And, if so, what classes of phenomena should be included as teleological, according to this more

1

basic understanding?

The steps in this program can be clarified by citing the process by which other pre-scientific notions have been developed into precise, scientific concepts. For example, biologists still divide organisms into plants and animals, even though this distinction originated with loose, ostensively defined notions having a slightly different denotation than they have for modern biologists.

Aristotle advanced our understanding of the plant-animal distinction by defining "animal" in terms of the faculties of locomotion and sensation. Contemporary biologists, possessing a more advanced knowledge about living organisms, use a more fundamental definition of "animal," one based on such characteristics as the type of metabolism (heterotrophic metabolism) by which animals carry on the basic life-process of self-maintenance.

The contemporary definition is more fundamental than Aristotle's in that the nature of an organism's metabolism explains more about that organism than does its ability or inability to locomote or sense. Organisms lacking locomotion and sensation, but possessing the animal, rather than the plant form of metabolism (e.g., the sponges), are more closely related phylogenetically, morphologically, and functionally to animals than to plants. Furthermore, it is the animal's dependence upon heterotrophic metabolism which makes both locomotion and sentience necessary: locomotion is a consequence of (among other factors) the need of heterotrophic organisms to locate other organisms upon which to feed; sentience is necessary to guide the animal in its movements through the environment.

Although the contemporary understanding of "plant" and "animal" makes the same general distinction as before, the contemporary definitions entail the inclusion of a somewhat wider class of organisms as animals than did the older definitions (e.g., sponges, corals, and protozoa are now included).

This illustrates the general process by which a loose, pre-scientific concept can be expanded to denote a wider class of phenomena than it originally denoted. Such an expansion is

2

prompted by the discovery of a new characteristic(s) which is more fundamental than those previously taken as defining. When that happens, it may be that some existents previously excluded from the class are now included, since they satisfy the new definition and are henceforth included in the class (see *Introduction to Objectivist Epistemology*, chapter 5).

My thesis is that certain teleological concepts can be expanded in this manner. Specifically, I will argue that the concept of "goal-directed action" can be defined in terms of properties more fundamental than those unique to its original referents (man's purposeful actions towards pre-envisioned ends), and that this new definition justifies the classification of all levels of living action—whether purposeful or automatic—as goal-directed.

I will also argue that the proposed definition of "goal-directed action" is not satisfied by inanimate processes. Goal-directed action depends upon a type of causation arising from the distinctive nature of life.

In short, I will show that men, animals, and plants act for the sake of obtaining certain ends, but rocks, rivers, and machines do not.

I
TYPES OF PROCESSES

THE ORIGIN OF TELEOLOGICAL CONCEPTS

Despite the technical nature of some teleological concepts, such as "goal-directed action," the use of teleological terms to describe various natural processes is a commonplace. English has such common teleological concepts as "goal," "purpose," "end," "aim," etc., and besides these there are many teleological phrases as well. We say, for example: "He studied diligently *in order to* pass the examination"; "He placed the dictionary on the desk *so that* it would be within easy reach"; "He takes daily walks *for the sake of* his health." (In each case the teleological phrase could be replaced by a teleological concept: "The goal of his diligent study was to pass the examination"; "His purpose in placing the dictionary on the desk was to have it within easy reach"; "The aim of his walks is to maintain his health.")

In fact, it is only at a relatively advanced stage of knowledge that some natural processes are viewed from anything other than a teleological perspective. Both primitive man and young children adopt an animistic attitude toward nature. They view all natural processes—both living and inanimate (in our terms)—as the expression of some conscious purpose. When a river rises, the primitive man explains this event in terms of the *will* of the river or of the river-god; when a storm occurs, its cause is held to be the anger of the clouds or of the cloud-god; a rock falling down a hillside may be seen as striving to get closer to the valley where it will "feel more at home."[1]

This animism is a natural consequence of a primitive stage of knowledge. We are directly familiar with human behavior, and it is only natural to assume that everything else in the world operates according to the same type of causation. The primitive man is aware that when people are angry, they make loud noises, violent movements, and become menacing. When, at this rudimentary level of understanding, he witnesses a thunderstorm, he assumes that the storm expresses the anger of the clouds. At this stage, he simply does not know enough about the nature of clouds and of man to appreciate the vast differences in their forms of action.

Even before primitive men have grasped the concept of "purpose" explicitly, they project the purposeful character of their own behavior out into the world. It is only as their knowledge advances that men grasp the idea of a cause that is *not* psychological in origin—or, more precisely, that they draw the distinction between psychological and physical causation.

The idea of *physical* causality—of actions caused by the physical nature of an entity without the intervention of conscious purpose—was not given explicit statement until the Greeks. The significance of the statement attributed to Thales, "Everything is water," lies partly in the fact that it is the earliest recorded expression of a naturalistic, non-animistic conception of the world.[2]

It seems, then, that teleological causation is the first form of causation to be understood. Teleological concepts originally derive from the experience of purposefulness which attends our own conscious, deliberate actions. These teleological concepts are then *secondarily* applied (correctly or incorrectly) to other kinds of processes. In applying teleological language to other entities, one is holding, by implication: these entities operate on essentially the same terms as man does. Thus, teleological concepts have as their original and paradigm case the purposeful behavior of man; it is a man's knowledge of his own purposefulness that gives rise to the wider idea of teleological causation.

LIVING ACTION VS. INANIMATE PROCESSES

As man's knowledge about nature grows, as he becomes increasingly aware of the differences between human purposeful action and other natural processes, the area to which he attributes teleological causation narrows. Historically, the first step in the movement away from animism probably came with the recognition that some entities in the world are alive and others are not. The distinction between the living and the non-living may have had its origin in the striking difference between the living and the dead, as Hans Jonas has suggested.[3] The most conspicuous feature of living organisms which isolates them as a class from both the dead and the inanimate is their power of self-produced movement. The two vital processes of growth and reproduction each involve self-produced movement. Dead bodies and other inanimate objects do not *move themselves* in this sense. (This property of living behavior will be discussed in detail in chapter 4.)

In recognition of the activeness of living behavior, I will apply the term "action" only to the self-produced movements of living organisms. This usage is somewhat stipulative; we often refer to inanimate processes as "actions" (e.g., the *action* of the moon's gravitational field upon the earth). It will, however, simplify matters to use "process" and "behavior" as wide, neutral terms to designate any relatively extended change involving the parts of a whole, and to reserve "action" for those processes which are self-produced.

Another readily observable distinguishing feature of life is its *conditionality*. By "conditionality" I refer to the fact that the organism's existence is conditional upon its actions: the possibility of death is always present, and survival is a positive achievement that must be continuously won and re-won by the organism. Accordingly, living organisms have needs—i.e., conditions which must be satisfied by positive action on the part of the organism, if the organism is to remain in existence. Inanimate objects do not, in this sense, have needs; the existence of inanimate objects is not in general dependent upon their performance of specific courses of action (apparent counterexamples to this generalization will be considered in later chapters). It is readily apparent that living organisms will perish if they fail in their self-sustaining actions, whereas inanimate objects can ordinarily remain in existence in a state of total, inert passivity. Living organisms maintain themselves by a ceaseless process of action to derive needed energy and materials from the environment and utilize them for self-maintenance.

These two features, self-produced movement and conditionality, ground the early distinction between the living and the non-living.

VEGETATIVE VS. CONSCIOUS ACTION

Continuing the progression to a non-animistic world-view, a further distinction is made within the realm of the living between conscious and vegetative actions. This distinction must be discussed in some detail. By "conscious actions" I mean those actions of a living organism which are initiated and directed by the organism's consciousness. In contrast, by "vegetative actions" I mean those actions of a living organism which are *not* initiated and directed by the organism's consciousness. A clear example of a conscious action would be the action of a dog chasing a rabbit. The dog's action is initiated and directed by its perception of the rabbit. If the dog had not seen or smelled the rabbit, the chase would not have been initiated; the dog, once losing sight and scent of the rabbit cannot continue to follow the rabbit's path. A clear example of a vegetative action would be the healing of a wound in the human body. The process of wound-healing is clearly neither initiated nor directed by one's consciousness—the process will occur just as readily when one is asleep or unconscious.

Coordinated biological actions take place in organisms which do not have the faculty of consciousness at all (e.g., plants), and these are also vegetative actions. By "vegetative action" I will denote any self-produced living action not initiated and directed by the organism's consciousness, regardless of the biological level—biochemical, physiological, or behavioral—on which the action is performed. The term "vegetative" is drawn from the Aristotelian tradition (cf. Aristotle's "nutritive" functions). (In biology, the term "involuntary

7

actions" is often used for the non-conscious actions of animals, but it would be somewhat inappropriate to describe as "involuntary" cases, such as plant growth, where the very capacity for conscious action is absent.)

Several important clarifications must immediately be made. First, what do I mean by the term "consciousness" here? By "consciousness" I mean the state or faculty of awareness, the ability to be aware of reality. (This cannot be taken as a definition of "consciousness," since "consciousness" and "awareness" are synonyms.[4])

Of course, a clear distinction must be drawn between consciousness and self-consciousness. An animal, such as a bird, is aware of the world without being aware *that* it is aware. A bird visually perceives an object, or feels pain, but is not aware that what it is doing is perceiving or feeling; to hold that the bird is conscious of its environment is not to hold that it has a concept of its self or its faculty of awareness. Man not only is aware, but also is aware that he is aware: man is self-conscious.

Secondly, the term "conscious" is not to be confused with the term "volitional." In describing an action as conscious, no implication is made that the action is chosen by an act of free will or as the product of a process of conscious deliberation. For example, a bird's action of nest-building is a conscious action, but it is presumably the completely necessitated outcome of the bird's biological/psychological makeup: the bird responds automatically to the pushes and pulls of pleasure and pain, with no power to deliberate or to exercise free will. The bird acts without volitional choice and without any knowledge of the biological significance of its action—e.g., the bird does not know that what it is doing is building a nest or why doing so is necessary to its survival. But if the action is initiated and directed by such experiences as seeing twigs, the action is to be classified as conscious in order to distinguish it from such actions as digestion, which occur independently of its conscious experiences.

The task of classifying the various actions of living organisms as conscious or vegetative is properly the province of biology rather than philosophy. Since we have no direct access to the consciousness of other organisms, our judgment as to whether or not their actions are guided by consciousness has to be based on inference. A reasonable basis for this inference is the similarity of both the structure and the actions of other organisms to related conscious actions in human beings.

We know by direct introspection that certain of our actions are guided by our conscious experiences, and it is reasonable to assume that where the physiological structures associated with such conscious experiences in man are present in the lower organisms, those organisms are also conscious. The main physiological structures involved here are, of course, the sense organs and the nervous system. Since dogs, for example, have eyes and a developed brain with cortical and sub-cortical structures quite similar to those in man that are known to enable human vision, it is logical to conclude that dogs can see (whether or not their form of visual experience is the same as ours). Moreover, when another species possessing sense organs and a nervous system performs behavior that gives evidence of using information supplied by these sensory systems, this confirms our conclusion that the behavior represents conscious action. An obvious illustration is that dogs can locomote more efficiently in an illuminated environment than in a darkened one. Further evidence may be provided by the existence of behavior which is apparently concerned with gathering information about the environment, such as "orienting responses" (as when dogs perk up their ears and appear to attend to sounds).

Descartes and others have held that none of this goes to *prove* that any organism other than man is conscious. It is conceivable, they say, that the information utilization of other organisms is non-conscious, just as an electronic computer can utilize information without being conscious. One can only respond that a standard of proof according to which "Dogs are conscious" counts as unproved is thereby invalidated.

I assume that consciousness is a faculty that has developed more or less continuously in evolution, like other biological capacities, from primitive forms in lower organisms to its most highly developed and complex form in man. Consciousness exists on a continuum ranging from perhaps the simplest organisms possessing a defined brain (i.e., the flatworms, *Planaria*) to the organism with the most developed nervous system: man. Naturally, it is a very difficult task to mark any one point on the phylogenetic scale at which consciousness first emerges. There is no basis, on the other hand, for positing the existence of consciousness in organisms of the plant kingdom. Plants lack both the structural and behavioral similarities to man which are the basis for inferring the presence of consciousness.

Accordingly, the *actions* performed by living organisms can be placed on a continuum of increasing complexity. At the beginning of this continuum we may place the simple one-step physical and chemical changes which are the components of coordinated biochemical processes. Oxygen exchange in the lungs of man, for example, operates by passive diffusion according to osmosis. The simple chemical reaction in which adenosine triphosphate (ATP) is converted into adenosine diphosphate (ADP) is another example of processes in this first category. (As will be argued in a later chapter, such simple physical and chemical changes, although constituting the components of living actions, do not in themselves qualify either as living actions or as teleological.)

At a slightly higher level of complexity are the coordinated, many-step biochemical processes of cellular metabolism. Examples are photosynthesis, glycolysis, and protein synthesis. Further along the continuum we reach the gross actions of plants, such as root growth, flowering, and phototropism. These involve the coordination of actions in several cells.

Proceeding to the animal kingdom, we come to the class of physiological processes, such as digestion, blood circulation, and kidney filtration. These may be regulated by lower brain centers, and are normally vegetative, since they can be performed during sleep or even under anesthesia.

The most complex actions that can still be classified as vegetative are the spinal reflex actions. Any reflex that can occur without requiring that the subject be conscious would qualify as vegetative. Certain posture-maintaining reflexes have been demonstrated to exist in cats whose spinal cord has been cut just below the level of the cerebrum.

The most primitive type of action that probably qualifies as "conscious" is what used to be called "instinctive behavior." Progressively, the somewhat mystical term "instinct" has been replaced in the literature by the term "stereotyped behavior." The distinctive features of such behavior are that it requires some perceived stimulus to serve as its "trigger" or "releaser," and that the behavioral sequence seems to be largely or entirely innate and resistant to modification within the life-span of an individual animal. Typical examples are the web-building behavior of spiders, the migratory behavior of birds, and the reproductive behavior of fish (see the classic work of Niko Tinbergen on the stickleback fish[5]). Stereotyped behavior seems to be triggered and guided by the animal's perception of features in its environment—and hence is conscious in my usage—but is distinguished from more complex types of conscious behavior by the fact that the "programming" for the execution of the behavior appears to be innate and relatively inflexible. For instance, if a spider's web in the midst of construction is partially destroyed, the spider will abandon it and begin again from the beginning, rather than repair the damaged sections and proceed on the same web, indicating that the whole sequence is pre-programmed.

A higher level of conscious action is constituted by learned behavior. Such behavior is not merely triggered and guided by sensory stimuli, but also is developed through the animal's conscious experiences and is, within limits, modifiable to conform to differing environmental conditions. Here again, the distinction between stereotyped and learned behavior is one of degree such that many borderline cases exist.[6] Cats, apparently, must learn how to kill their prey; certain maternal behavior in female rats is dependent on contact with other rats; and all cases of "instrumental conditioning" require the learned association of new behavioral sequences with reward and punishment contingencies.

11

Finally, at the end of the continuum, there is the rational behavior of man. A man's action of writing a book, choosing a career, or simply cooking a meal depends on his use of abstractions, i.e., concepts.

This is the continuum of organic behavior. It is a *continuum* in that we are confronted with different ranges along a single basic axis: degree of complexity (and integration) of the action involved. As stated previously, the ranking of types of behavior along this continuum is a problem for biology, not philosophy, since it depends on the factual nature of the behavior. Although I have suggested that the line of division between vegetative and conscious action is to be drawn between reflexes and stereotyped behavior, respectively, my overall thesis is independent of that position. All that is assumed is that human actions are conscious and that some other living actions are vegetative—exactly which other actions count as vegetative affects only the choice of examples, not the basic thesis.

It may also be noted that in addition to borderline cases and controversial cases, certain types of action may change their status: some conscious actions may with repetition approach the vegetative, and certain normally vegetative actions may with training come under conscious control. For instance, an infant has to learn to stand upright by a conscious process, but after much practice, the maintenance of a standing posture becomes automatized, so that progressively less conscious attention is required. For an adult, standing erect seems to be less than purposeful but more than vegetative. An adult in the midst of an engrossing discussion may maintain his posture without giving it any (or hardly any) conscious attention, yet to remain standing, he has to make continual muscular adjustments requiring sensory input; he cannot stand while anesthetized. Conversely, the heartbeat is an action which in human beings is normally vegetative; some individuals, however, have learned to control their heartrate consciously. The process of breathing in man is an example of an action which can occur either consciously (as in swimming) or vegetatively (as in sleep). Even spinal reflexes, such as the patellar reflex in man, are modified by the intervention of higher brain centers associated with conscious states: a man who is tense or anxious will tend to show an exaggerated patellar reflex. Behavior which is normally vegetative in one species may be normally conscious in others, and vice versa.

The conscious/vegetative distinction, then, does not mark out two inflexible and mutually exclusive categories such that every action can be classified unambiguously as one or the other, out of context. We are dealing not only with a continuous spectrum, but with a spectrum whose contents exhibit a certain plasticity. Whether an item of behavior is conscious or vegetative depends on what species of organism is being considered and on the particular situation and the particular history of the individual within that species.

Finally, a conscious action always involves sub-processes which are vegetative in terms of their own operation. The gross action of walking is conscious, but the contraction of individual muscle fibers in the gastrocnemius muscles of one's legs would be classified as vegetative actions. Another example is the action of visual perception. The percept which results from this process is sometimes achieved consciously, as in the case of a man who is purposefully looking at something. The neurophysiological processes which underlie the perception are not, in themselves, conscious. When, for example, a man is intently looking at a painting, the looking is conscious, but the release of neural transmitters in his optic nerve is vegetative.

The forgoing problems and complexities relate to the *application* of the conscious/vegetative distinction and consequently need not concern us. The validity of the distinction itself is undeniable. Certainly my action of writing this paragraph is initiated and directed by my consciousness, but the metabolic processes currently going on in a cell in my finger are not.

GOAL - DIRECTED ACTION VS. PURPOSEFUL ACTION

Although primitive man regards all processes in nature as

similar to his own purposeful actions, modern man distinguishes firstly all living actions from inanimate processes, and secondly conscious from vegetative living actions. How do these distinctions affect the area of behavior one considers to be teleological?

A conscious action can be, under certain circumstances, teleological: our own purposeful actions fall into this category. As we have seen, human purposeful actions are in fact the original source of our teleological concepts.

The precise features constituting the purposefulness of man's conscious actions will be identified in chapter 3. At this point, it is necessary only to note that I take the term "purposeful" to be much narrower than some other teleological concepts, such as "goal-directed." I am assuming that "purpose" and "purposeful" apply only in the case of conscious actions—that is, in the case of actions in which the agent has an "end-in-view." According to this usage, a "non-conscious purpose" is a contradiction in terms.[7] This usage, while somewhat stipulative, accords with the *Oxford English Dictionary* ("Purpose: 1. The object one has in view."). The origin of the term is the Latin *pro*—before, and *poser*—to place, a purpose being something one places before himself—not, of course, placed physically in front of oneself, but projected mentally ahead as a future goal.

(For the adjectival form, I use the term "purposeful" rather than "purposive." "Purposive" can be read as "purpose-like," while what I want to designate is "purpose-having"—i.e., cases in which the agent has a conscious purpose.)

The teleological concepts derived from the case of our own purposeful action can be straightforwardly applied to some of the conscious actions of other animals. For instance, a dog eats *in order to* satisfy its hunger, or for other motives, such as to enjoy a food it happens to like. Of course, there are very important differences between the purposeful actions of man and those of the higher animals. When a man eats, he has an implicit or explicit *conceptual* awareness of what he is doing and of the end it is to serve.

Moreover, human ends, being abstract, are much more

complex than those of a higher animal. One may eat in order to maintain his health, to please a host, to sample a new food, or as part of an experiment in nutrition. To say that a dog eats purposefully to satisfy its hunger, we need recognize only that the dog feels hungry (without having the conceptual identification of that state as hunger), sees or smells the food (without having the conceptual identification of it as food), associates the sight or the scent with the pleasure of satisfying hunger (again without having conceptualized knowledge of the relationships involved, but merely a perceptual association based on memory), and goes to eat the food.

Accordingly, even though the *level* of conscious direction and motivation differs profoundly in the action of the man and the dog, both actions qualify as "purposeful" in the sense of that term I am using. The differences in the types of conscious action involved warrants a distinction between conceptual purposes and perceptual purposes.

In other words, there is no objection to talking of teleological ends, goals, etc., where there are, in Dewey's phrase, "ends-in-view"—whether this "view" is perceptual or conceptual. Again, the factual question of just which other organisms have conscious purposes is not the issue: *if* an organism acts toward an "end-in-view," no special philosophic problem arises in describing that action as teleological.

At the other extreme are the inanimate processes occurring in nature. Once a natural process is recognized as inanimate, it is regarded as non-teleological. Once it is realized that the storm clouds, for example, are purely physical phenomena, governed neither by life nor mind, the teleological explanation of the thunderstorm is abandoned. To recognize a natural process as inanimate is to see it as so different from one's own purposeful actions that the application of teleological concepts to it becomes absurd. (If someone insisted on using teleological terms, such as "goal-directed," to include all natural processes, he would then need to invent new terms to form a distinct class for those processes essentially similar in their mode of causation to our own conscious actions. The issue is not linguistic, but factual: our categorization must conform to the similarities and differences in the subject matter.)

It is specifically with the class of vegetative actions that the problem of teleology comes into sharp focus. The other two kinds of process have been preliminarily classified: conscious action can be teleological; inanimate natural processes, being clearly different in their mode of causation, are classified as non-teleological. The classification of vegetative actions, however, is not easily settled. Are the vegetative actions of living organisms teleological? Are vegetative actions closer in their mode of causation to inanimate processes or to purposeful voluntary actions? This is the central issue, and its proper resolution will provide the basis for a new and deeper understanding of teleology.

II
ALTERNATIVE POSITIONS ON VEGETATIVE ACTION

There are basically two positions concerning the classification of vegetative action: either it is teleological or it is not. But after that simple dichotomy, the schools of thought branch out, and the issue becomes somewhat more complex. In outlining the array of possible positions, some terminological simplification needs to be introduced. I will refer to the two basic schools as the "teleologists" and the "mechanists" according to whether or not they hold that teleological concepts can properly be applied to the vegetative actions of living organisms. (Please note that I am not referring to the sense of "mechanism" in which it opposes "vitalism" on the question of whether or not vegetative actions transcend physical-chemical law.)

THE MECHANISTS

The position of those I am calling the "mechanists" is fairly straightforward. They hold that despite the *apparent* purposefulness of vegetative actions, such actions are non-teleological. Vegetative actions, they maintain, are too different from purposeful actions to be included in the same causal category. The mechanists hold that there is no fundamental common feature that could serve to isolate vegetative actions as a class from the ordinary mechanical processes of inanimate matter with regard to the mode of causation involved. (Obviously, there is the common property of being *living* rather than inanimate, but we are interested here in classification with respect to the issue of teleological causation.) According to the mechanists, the application of teleological concepts to vegetative actions, such as plant growth, is improper because it obscures what they regard as the more important differences between vegetative actions and con-

17

scious actions.

The mechanists characteristically adopt the Cartesian position that the fundamental division of all natural processes is between those under the control of mind and those that are not—or, in our terms, between conscious actions and all other processes in the universe. As a secondary issue, non-conscious processes can be subdivided into the living and the inanimate—but mechanists deny that this can be the primary distinction (see figure 1). The mechanists hold that the application of teleological descriptions and explanations to vegetative living actions would carry the unwarranted implication that such processes are under the control of a purposeful mind. For this school, in short, the only genuine ends are ends-in-view.

THE TELEOLOGISTS

The teleologists, on the other hand, maintain that there is a fundamental mode of causation common to all types of living action, whether conscious or non-conscious: living action is *goal-directed*. (The term "goal-directed action" was popularized in E. S. Russell's work, *The Directiveness of Organic Activities*; my adoption of this term does not imply any agreement with Russell's views.) According to the teleologists' conceptual system, although the family of teleological concepts originally derives from the case of conscious, purposeful action, many of them, such as "goal," may be validly applied to vegetative actions as well, on the grounds of the fundamental similarity they see as uniting vegetative and conscious action in this regard.

According to the teleologist view—the view to be defended in this book—purposeful action is just one form, or subcategory, of goal-directed action. In purposeful action, the goal of the process is held in conscious form, and it is some mental content which directs the subsequent action toward the goal. In vegetative action, the teleologists maintain, despite the fact that the action is automatic and non-conscious, there is still a genuine goal that in some sense directs the actions by which the organism attains that goal.

Figure 1.

MECHANIST PERSPECTIVE

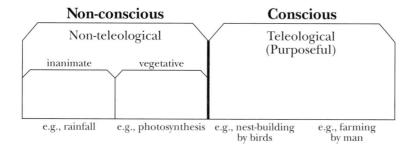

Figure 2.

TELEOLOGIST PERSPECTIVE

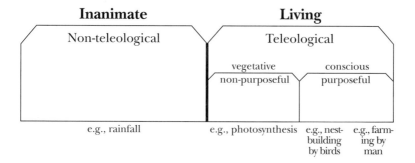

Teleologists differ, of course, as to just what is the fundamental similarity uniting vegetative and conscious action—i.e., they differ as to what constitutes goal-directedness. But *qua* teleologists they agree that the basic division of natural processes is between the animate and the inanimate, and that animate actions are teleological but inanimate processes are not (see figure 2). Teleologists do not hold consciousness to be a necessary component of teleological action — there can be ends which are not ends-in-view. An action such as the growth of a plant is held to be goal-directed despite the absence of any conscious purpose on the part of the plant (and despite the absence of any "Divine design" in nature).

The teleologists divide into two categories with regard to their views on the metaphysical status of goal-directed action. I will term these two views the "vitalist" and the "emergentist" schools.

The vitalist teleologist views the distinction between the teleological and the mechanical as an absolute dichotomy consisting of two distinct, mutually exclusive principles of action. According to vitalist teleology, goal-directed action is a property *sui generis* and irreducible; goal-directed action is, by its very nature, non-mechanical or even anti-mechanical. Just as the general vitalist position in biology holds that *life* transcends the laws governing inanimate matter, so the vitalist position regarding teleology holds that *goal-directedness* transcends the laws governing mechanical processes. The Austrian biologist Schubert-Soldern, an avowed vitalist, states:

> The living organism exhibits chemical processes which tend in a direction of which the molecules by themselves are quite incapable. . . . If the matter in the living organism is taken over functionally into a radically different sphere of being, this does not mean that it loses any of its special character as matter. It means that something is superimposed on it. The entelechial life-principle seizes upon matter, takes possession of it and determines it in a special way.[1]

The *emergentist* teleologist, on the other hand, sees goal-directedness as an "emergent property" of matter. By an "emergent property" I mean a property which is possessed by the whole *qua* whole and is not possessed by the parts.

One must clearly distinguish this view from others which go by the same name. There are a number of different positions in various areas of philosophy which are called "emergentist" (see Ernest Nagel's discussion in chapter 11 of *The Structure of Science*). What I am here calling the emergentist school of teleology is *not* characterized by C. D. Broad's view that an emergent property is a property of a whole which could not be *predicted* from a complete knowledge of the properties of its components. Emergentism, in my terminology, is a metaphysical rather than an epistemological position. This view of emergentism is well expressed by biologists Robert Haynes and Philip Hanawalt: "insofar as any complex system is more than the sum of its parts, it is so by virtue of the controlled interactions or modes of regulation of these parts."[2]

According to this emergentist position (which I will be defending), teleological causation on the vegetative level is not an *alternative* to mechanical causation—rather, teleological causation is a complex *form of* mechanical causation. Thus for the emergentist teleologist, vegetative action is *both* mechanical and teleological, in contrast to the inanimate processes which are only mechanical. According to the emergentist conception, the mutually exclusive terms are not "mechanical" and "teleological," but rather "*merely* mechanical" and "teleological-mechanical." The movement of a rock falling down a hill, for instance, is merely mechanical—there is no teleological dimension to the process; the action of cell division, on the other hand, is, for the emergentist teleologist, both mechanical and teleological.

Analogously, in physics, even though there is a distinction between atoms and molecules, we recognize that molecules are composed of atoms, and that the proper dichotomy is between separate, dissociated atoms and atoms combined into molecules. The equivalent of the vitalist position in this context would be the idea that the actions of a molecule are not simply a function of the individual properties of its constituent atoms in the specific organization dictated by those properties, but that there must be some "molecular entelechy" which is superimposed on the atoms, causing them to act in ways which transcend the individual properties of the atoms themselves.

Of course we must be clear that the emergentist is not denying that the whole may exhibit properties different from those exhibited by its parts—just the opposite. The emergentist maintains that when the whole exhibits new properties, these new properties are the result of the integration of the parts into the whole.

The opponents of emergentism always contrast the action of the parts within the whole to the action of the parts when dissociated from one another, then they attribute the difference in behavior to some independent "principle of order" or "entelechy." Schubert-Soldern, for instance, contrasts chemical processes occurring in a living organism with those exhibited by the "molecules by themselves"—i.e., alone, outside of an organism. But for the emergentist, this difference between parts in isolation and parts interacting is a consequence solely of the nature and arrangement of the parts themselves, without requiring any outside principle or force.

To take a very simple illustration, imagine we are presented with two hemispherical pieces of wood, each having a sticky substance on its flat side. Alone, neither hemisphere can roll; when joined to form a sphere, the whole can roll. Rolling is thus an emergent form of action completely determined by the individual separate properties of the parts and their arrangement. Obviously the whole formed by uniting the two hemispheres is, in a sense, "greater than its parts"—but it is just as obvious that this "extra something" of the whole (its ability to roll) is not to be explained by the supervention of a "principle of order" or "entelechy." There is no "transcendence" of the natures of the parts nor of the laws governing their behavior.

How, then, does the emergentist teleologist differ from the mechanist? The emergentist teleologist holds that the mechanical action of living organisms is, due to its special complexity, goal-directed. The mechanist holds that no matter how complex the process is, it cannot be termed "goal-directed" unless conscious purposes are involved. Thus the emergentist teleologist and the mechanist agree on the metaphysical status of vegetative action, but disagree about its goal-directedness. The emergentist teleologist and the vitalist teleologist agree that vegetative action is goal-directed, but disagree about the metaphysical status of that property.

Finally, we must distinguish within the emergentist teleologist school two views concerning the cognitive importance of teleology. One view holds that since goal-directed action is not, as the vitalists claim, "transcendent," since it is "only" a complex form of mechanical action, teleological concepts are just excess baggage that should be ignored or eliminated. The other view—the view I will be defending—maintains that even though terms such as "goal-directed action" *are* definable in non-teleological terms, this does not mean we should dispense with teleology and teleological concepts. Teleological concepts, in my view, refer to actually existing emergent properties, and it is proper to recognize these properties in our accounts of living action.

In general, the fact that a given concept can be defined via other more basic terms does not mean that the concept is "nothing but" a dispensable abbreviation for those more basic terms. A term does not equal its definition. A good example is provided by the term "war." "War" is a global concept which, in any given instance, is reducible to the actions of individual men in personal combat (plus the individual actions of other men acting as commanders, strategists, and statesmen). No one supposes that a war involves anything which "transcends" the interrelated actions of the individuals involved—there is no appeal here to the equivalent of an "entelechy." Nonetheless, the concept "war" is not a useless redundancy; there is an obvious cognitive need for such a concept. How, in the absence of the concept "war," could one hold in one's mind such a proposition as: "The harshness of the terms of surrender which ended World War I was a factor in causing World War II"?

The position adopted in this book is that explanation on the level of parts does not necessarily eliminate the need for explanation on the level of wholes, and vice versa. Both approaches contribute to our understanding of complex phenomena. Evolutionary biologist George G. Simpson argues convincingly for this position in his work *This View of Life*.

In biology, then, a second kind of explanation must be

> added to the first or reductionist explanation made in terms of physical, chemical and mechanical principles. This second form of explanation, which can be called compositionist in contrast with reductionist, is in terms of the adaptive usefulness of structures and processes to the whole organism. . . . It is still scientifically meaningful to say that, for instance, a lion has its thoroughgoing adaptations to predation *because* they maintain the life of the lion.[3]

Accordingly, the teleologist position to be defended here may be described as "non-eliminativist emergentism." That is, I will argue that in fact there is an emergent phenomenon—goal-directed action—which exists in two forms: the purposeful and the non-purposeful, according to whether the action is conscious or vegetative, and that biologists can and should employ both mechanical and teleological explanations for living actions.

OTHER SCHOOLS

Before proceeding, mention must be made of two other schools, which are often referred to as teleologist, but which are essentially different from the position I will be defending.

First there are those who attribute consciousness to all levels of life, and argue that all living actions are teleological in that they are all guided by conscious purposes. In the view of this school, no such category as "vegetative action" exists, since *all* living action is held to be under conscious direction. If, however, one recognizes that only certain higher processes in animals are conscious, the position of this school is mechanist by implication, because its adherents hold that only conscious purposes can qualify as teleological. They differ from the ordinary mechanists only concerning the factual question of where to draw the line between conscious and vegetative actions—this school admitting more processes as voluntary than would the ordinary mechanist. They both hold the same essential premise, however: the only genuine ends are ends-in-view. Both refuse to accept the validity of the concept of goal-directed action as denoting a class of processes *wider* than purposeful actions. Edmund Sinnott is a clear representative of this school:

> The position which I propose to defend—the thesis I am nailing to the cathedral door—is briefly this: that biological organization (concerned chiefly with organic development and physiological activity) and psychical activity *are fundamentally the same thing.* This may be looked at from the outside, objectively, in the laboratory as a biological fact; or from the inside, subjectively, as the direct experience of desire or purpose. . . . To talk about "mind" in a bean plant or a protozoan, or even in a worm, may seem absurd, but it is more defensible than trying to place an arbitrary point on the evolutionary scale where mind, in some mysterious manner, made its appearance. We are dealing here with a quantitative and not a qualitative difference.[4]

This view, though outwardly super-teleological, is in fact compatible with the position of the ordinary mechanist, as Sinnott recognizes:

> An advantage of the hypothesis here presented is that it can be accepted by either the materialist or the idealist as a sound interpretation of purposiveness. It does not take sides. It implies neither mechanism nor teleology, fate nor freedom, but simply attempts to tie together, as identical, the biological and the psychological events.[5]

The second view from which I wish to dissociate my own position is one which attempts to straddle the fence between the Sinnott-type approach and ordinary mechanism. This school holds that while vegetative actions are not under conscious direction, they are teleological because they are guided by something *like* a conscious purpose—an "urge," or "striving," or "conatus." This "conatus school," as I call it, never succeeds in giving any content to the very notion of "conatus" (or the like) on which it depends. All we are told about the "conatus" is that it plays the role that foresight or desire plays

in the case of purposeful action, but that it is "pre-conscious." Thus the idea of unconscious foresight or unconscious desire is actually nothing but an "x-factor" posited solely to support their theory.

The conatus theory of vegetative action, then, faces a dilemma: either the "conatus" is a state of consciousness or it is not—if it is a state of consciousness, then the conatus theory collapses into the previous view that all life involves consciousness, and it thus abandons the wider concept of goal-directed action (plus raising the criticism that the attribution of consciousness to such processes as plant growth is arbitrary)—if, on the other hand, the "conatus" is truly non-conscious, we are left in the dark as to what it *is*. In short, the conatus school stands convicted of positing the existence of a phenomenon defined in terms of apparent contradictions (e.g., "unconscious desire") and for which no independent evidence is available.

Some mechanists do not oppose the use of teleological concepts to describe vegetative action. According to these thinkers, the issue is merely one of semantics, of finding the most "convenient" or "fruitful" terminology, rather than being an issue of fact. It is important to realize that this attitude already presupposes the mechanist position, and hence cannot be used to forestall the need of deciding which view of vegetative action—mechanist or teleologist—is correct. For if there is indeed an emergent goal-directedness in vegetative action, the decision of whether to use teleological language is surely not an optional one. The issue is optional only to those who are already convinced that vegetative actions do *not* have a teleological character which sets them off from the purely mechanical processes of inanimate matter. To hold that the issue is non-factual is to have already decided that biological teleology is not a fact.

In a similar vein, the "cyberneticists" led by Norbert Wiener, have attempted to salvage teleological language by defining such concepts as "goal" in terms of "negative feedback" or similar "input/output" relations. These definitions, however, are so general that they render many, if not all, inanimate processes teleological (see chapter 5). This approach is unjustified if there is a more fundamental sense of "goal-directedness" than those given in behavioristic definitions, a sense which is satisfied by living organisms and living organisms only. The cyberneticist approach, then, assumes in advance that no such exclusively biological teleology can be demonstrated.

METHODOLOGY

The burden of proof for demonstrating the existence of an exclusively biological teleology rests with the teleologist. In defending this view, it is up to me to show that there is an objective basis for extending the teleological concepts derived from purposeful action to include vegetative action as well. I must show that there is a new level of causation which fundamentally differentiates living action from inanimate processes. I must provide a definition of "goal-directed action" which explains, in objective, scientific terms, how teleological causation operates.

How is this to be accomplished?

In order to decide whether vegetative action is teleological, we need to have some antecedent idea of what teleological causation is. Otherwise, the question would be equivalent to: "Is vegetative action glyxical"—where we have no idea what "glyxical" means. As I have stressed, we get our idea of teleological causation from considering our own conscious behavior. We first have the idea of teleological causation when we recognize such behavior as being *purposeful.* Hence the question "Is vegetative action teleological?" amounts to the question "Is vegetative essentially similar in its mode of causation to purposeful action?" Therefore, the first thing we need to know is just what is the mode of causation in purposeful action. Then we need to analyze the kind of causation exhibited in vegetative action: how is the causation of vegetative action similar to that of purposeful action and different from that of inanimate processes? Finally, we must assess the results: do the similarities between the causation of vegetative and purposeful action warrant the description of both as "goal-directed" in contrast to the purely mechanical causation of inanimate processes? This, in outline, is the procedure I will employ to demonstrate the goal-directedness of living action.

Accordingly, the positive part of this book will consist of answering three questions:

> 1. What are the properties of conscious action which make us view it as purposeful?
> 2. In what ways does vegetative action exhibit properties similar to those constituting the purposefulness of conscious action?
> 3. Are these similarities mere analogies between two fundamentally different kinds of action, or do they imply that vegetative and purposeful action are two sub-categories within a single general type of causation: action directed by a goal?

These same steps characterize the procedure used in other areas. For example, geometry students are normally familiar with the concept of "rectangle" before they have learned the concept of "parallelogram." It is possible to show them, however, that the rectangle is actually just a special case of the parallelogram (just as I will argue that purposeful action is a special case of goal-directed action). This would be achieved by following a procedure identical to the one outlined above. One would begin by identifying what are the properties of figures such as ▭ and ▯ which makes us say they are rectangles. These properties would be: having four sides of which two pair are parallel and having four right angles. Next one would identify the properties of such figures as ▱ and ▱ which are similar to those of the rectangle. These properties are: having four sides of which two pair are parallel and having two equal acute angles and two equal obtuse angles. Finally, one can show that the similarities between the two types of figures are fundamental and that we can describe both as "parallelograms," defined as: a four-sided figure with two pairs of parallel sides and two pairs of equal angles.[6] This can be done by showing that the distinctive property of the rectangle—having four right angles—is subsumed as a special case under the wider property of having two pairs of equal angles (since four right angles may be viewed as two pairs of right angles). Thus one would have shown that figures such as ▢ and ▱ are not merely analogous, but rather that their observable similarity (as opposed to figures such as △) results from their common possession of the properties defining the parallelogram. In the same way, I will show that purposeful action and vegetative action are not merely analogous, but rather that their observable similarity results from their common possession of the properties constituting goal-directed action.

III

THE ANALYSIS OF PURPOSEFUL ACTION

"Purposeful" is a teleological concept—perhaps the primary teleological concept. The investigation of teleology should begin, therefore, with the question: What features of man's conscious actions gave rise to the concept "purposeful"?

It is impossible to conceptualize a phenomenon to which no alternative has been encountered or imagined. For example, the state of being awake can be conceptualized only by contrast to the state of being asleep. A being who knew nothing of sleep could not grasp the concept "awake." Likewise, in order to recognize explicitly the purposefulness of one's own actions, one must differentiate it from some contrasting non-purposeful form of action.

ANTICIPATED VS. ACCIDENTAL CONSEQUENCES OF ACTION

One way to make this contrast is by differentiating actions done purposefully from actions done accidentally. Even at a very primitive state of knowledge one can observe that one's actions have two very different kinds of consequences: those which one *intended* and those which one did not.

For example, suppose a man walking to some distant location notices that he is leaving a trail of footprints behind him. He did not intend to leave the footprints: that was not his purpose; his purpose was to reach his destination. The footprints were made by accident.

Generalizing from simple examples of this kind, one can distinguish the intended consequences of one's actions from accidental by-products. What is the basis of this distinction as grasped so far? The intended consequences are, first of all, those which the agent *foresees* or *anticipates*. In the preceding example, the man was not at first aware of the footprints he was making, but he was of course aware that by walking he was approaching his destination.

Thus, one characteristic which distinguishes purposeful action from accidental action is that the former involves the agent's awareness or mental anticipation of some consequence of the action he performs. A man who is not aware that his walk will bring him to a given spot could not be said to act purposefully to reach that spot. The agent's anticipation that some consequence will occur is a necessary condition for the action to be purposefully directed toward the attainment of that consequence. But this necessary condition is not sufficient to define "purposefulness," for a purposeful action may have consequences which, though anticipated by the agent, are wholly irrelevant to the purpose he is seeking.

For example, once the man realizes that his walk will leave footprints, it is not necessarily the case that leaving footprints becomes part of his purpose, or that he leaves them purposefully. We would say that leaving the footprints was not his reason for taking the walk (nor even part of his reason). In order that a consequence of one's action qualify as part of one's purpose for taking the action, it must be the case that one's anticipation of effecting that consequence is the *cause* of one's performance of the action. (Similarly, for a consequence to form *part* of the purpose of a man's action, it must be part of the cause of his performance of that action.)

Of course, the accidental is not the causeless; when we say in this context that an event was *accidental*, we are indicating the absence of one particular kind of causal connection which might have been present. In speaking of the accidental consequences of a man's actions, we are ipso facto granting that they are caused ("consequences" are after all effects), but we are indicating that the prospect of bringing about these consequences was not a cause of his performance of that action. Something is brought about accidentally when the belief that it would be brought about was not what motivated the agent to undertake the action. A purposeful action, in contrast, involves *causation by* the agent's anticipation of some consequence of that action.

In any given case of purposeful action, the agent's anticipation that he will bring about the intended consequence may be either true or false, rational or irrational. There is no rational justification for a savage's belief that by making animal sacrifices to the gods, he will influence the weather. Nevertheless, we can certainly say that the purpose of such sacrifices is to influence the weather if it is his hope of influencing the weather that causes him to make the sacrifices. On the other hand, the consequent elimination of the sacrificed animal as a source of food is an accidental consequence of the action: the savage is not making the sacrifice *in order to* diminish his food supply, even though he may very well be aware that this will be a consequence of his action.

Furthermore, we need assign no specific degree to the agent's confidence in the efficacy of his action. The agent need not be *certain* that his action will be successful, nor even believe success likely; it is only necessary that the agent believe there is a possibility, however remote, that his action will be successful and that it is his belief in the possible efficacy of the action that causes him to undertake it.

For an action to be purposeful, then, it is necessary only that the agent undertake it because he believes it *can* bring about a given consequence, whether his belief amounts to a confident expectation of success or only the barest degree of hope. I use the word "anticipation" in a weak sense to cover all degrees of confidence in the efficacy of the action. If a person were thoroughly convinced that a given course of action could not possibly bring about a given result, but still undertook the action, whatever his purpose might be, it would *not* be an attempt to bring about the result in question.

THE ROLE OF DESIRE

There is one remaining characteristic essential to (normal cases of) purposeful action. To elucidate this characteristic we need only ask: What is the difference between those anticipated consequences which cause a man to act and those which do not? What is it about a projected consequence of a given action that causes a man to perform the action which he hopes will bring about that consequence? The short answer is: desire. In ordinary cases, it is one's desire for some anticipated consequence of an action that causes one to perform that action. Desire for some anticipated consequence of an action is certainly an integral part of the original concept of purposefulness.

Of course, there are purposeful human actions in which the element of desire is missing, as when a man chooses to perform an action out of a sense of duty against strong desires in the opposite direction. These cases, however, have no bearing on the issue at hand, since we are concerned only with paradigm cases of purposefulness, not with "hard cases." We are not interested in the specific complexities of human motivation, but rather in identifying what kinds of actions give rise to teleological concepts.

The use of the term "desire" to indicate the mental attitude of the agent toward his purpose is somewhat of an oversimplification. If "desire" is taken to mean a felt longing, it is often lacking in purposeful action. A man reaching for an ashtray may not be experiencing any particular emotion, though his action is clearly done on purpose. Nevertheless, in a loose sense, an observer could appropriately say that the man *wanted* to have the ashtray. In such cases, the low degree of conscious attention given to the object and the immediacy of the "desire's" satisfaction account for the absence of any definite emotion toward the goal or purpose. If, on the other hand, a man were to devote a great deal of effort over a long period of time to achieving a given result, yet he reported that he was emotionally indifferent to his success or failure, we would seek some special explanation of this puzzling behavior. We would wonder if he is emotionally repressed, if he is dissembling, etc.

Furthermore, I do not mean to imply that desire is an irreducible primary in human action. As an emotion, desire is an effect of more basic mental processes, such as appraisal and evaluation. The situation is not altered if purposeful action is analyzed in terms of the *value* that an agent places, consciously or subconsciously, on attaining a given state of affairs. It is to be understood that in discussing the role of "desire" in purposeful action, I actually mean either the emotion or the mental appraisal which underlies it.

We may summarize our analysis of purposeful action in the following statement:

A purposeful action is a conscious action caused by the agent's desire for some anticipated consequence of his action.

We have now arrived at an understanding of purposefulness which is adequate to our task: the identification of the source of teleological concepts. All our teleological concepts and expressions are derived from the case of purposeful action. The mode of causation in purposeful action has been described: a desire for some projected future consequence of an action causes the agent to perform that action. These are the properties of conscious actions which make us view them as purposeful.

The concept of purposeful action, understood in this manner, is applied without difficulty to the conscious actions of animals below man in the evolutionary scale. Take the case, for instance, of a hungry animal approaching a source of food. All the elements of purposeful action may be present, though in a less complex form than in the case of human purposes. Let us first consider the element of desire.

Of course, an animal does not desire, say, food in the clear, conceptual form in which man is capable of experiencing his desires. The animal has no *word* for food, eating, or hunger, nor does the animal grasp the abstractions which the words "food," "eating," and "hunger" denote for us. But the presence of a desire for an object does not require that one have a conceptual identification of the object or of the desire. The animal simply senses the food and associates this sensed object with the satisfaction of its hunger, causing it to have a felt impetus toward the food—i.e., a desire for the food.

Secondly, it can be said that some infra-human animals *anticipate* the satisfaction of such desire, if we interpret "anticipation" widely to include perceptual association. For example, in saying that a dog "anticipates" pleasure, one does not mean that the dog thinks to itself "If I walk to the food and eat it, I will feel pleasure." It is true, however, that the dog has learned, through experience, to associate eating with pleasure and that it associates walking toward the food with eating. It is in this sense that we may say the dog anticipates the satisfaction of its desire: it mentally associates this satisfaction with the performance of its action.

Clear evidence of the causal role played by an animal's anticipation is provided in the case in which it does not know what kind of action will satisfy a given desire. For example, a dog being trained to sit on command may very well desire to be rewarded by its master; it does not yet know, however, that the action of sitting will gain it this reward. The process of training a dog to sit on command is not one of teaching him how to assume the sitting position—the training is devoted to having the dog *associate* the command with the action and with the reward. Until the dog can associate the command with the action, it cannot *purposefully* act to obey the command. Canine obedience requires the dog's mental connection of the action, when performed under certain conditions, with its consequence. Psychologist Magda Arnold summarizes:

> In every formal learning situation there is a goal set by the experimenter or the person, to which the experimental subject has to find his way. In such learning, the present situation must be appraised as similar to a past one, promising a similar satisfaction or demanding a similar action.[1]

One might object that my characterization of purposefulness, since it involves causation by desire, implies the absence of free choice in man and animals. However, in saying that purposeful actions are *caused* by desire, I do not mean to imply that such actions need be *necessitated* by the agent's desire, nor that the agent is the helpless puppet of his emotional responses. What kind of causation am I then attributing to desire? The desire for the goal is that which causes the action *either* by directly necessitating the action, or (in cases of free choice) by the agent's choice to *allow* the desire (or the value-judgment underlying it) to govern his action. Only if one chooses to recognize and hold in full conscious awareness the value of a given course of action will he perform it. Thus, a potential goal becomes a cause of one's actions only through his volitional choice. In free-will choices, the agent endorses the desire and thus gives it causal efficacy. But in the absence of desire (or other positive appraisal), the action would not be performed.

The desire then, is a cause in the sense of contributing to the action's occurrence; this is true generally, for animals and man, even though an additional choice is required, in the human case, before the desire is acted upon.[2]

THE ORDER OF CAUSE AND EFFECT

This general analysis of purposeful action facilitates the resolution of a paradoxical feature of teleological causation: the appearance of a reversal of cause and effect. In teleological processes it *seems* that the final state, or end is a cause of the action directed toward the achievement of that end.

The view that the goal of an action can be a causative factor in the origination of that action is manifested in our teleological language itself. When we say an action was performed *in order to* obtain its goal, we mean that it is the requirements of achieving the goal that caused the agent to undertake the action. We often say that an action occurs *so that* a given consequence may be brought about, again with the implication that the consequence of the action causes the selective performance of just the kind of action which will have that consequence as its effect. The same thought is contained in the phrase "the end determines the means"—the implication is that the end exists prior to the action and causes the appropriate steps to be selected for performance. The very description of an action as a *means* implies the causal subordination of the action to its end.

Likewise, we *explain* those actions we regard as teleological by reference to their effects. We treat the effects as the causal conditions explaining the initiation and direction of the action. For example, a man drives to the railroad station in order to meet his wife. Why does he leave his home and travel in a certain direction? The explanation is given in terms of a future event which is the action's goal: meeting his wife.

If taken literally, the idea of *final* causation, of causation by a future end or goal, is absurd. How can an event which is to take place in the future be the cause of a present action? On a metaphysical level, the suggestion of a future event acting as the cause of a present action is contradictory: the future event does not exist yet, and what does not exist cannot act as a cause. On an epistemological level, explanation in terms of final causes (taken literally) presents us with a vicious circle: the action is to be explained by the future event, but the future event's realization is to be explained by the action.

The full absurdity of a literal reversal of cause and effect is dramatized in the case of unsuccessful action. Suppose the husband in the previous example arrives at the train station and discovers that his wife is not on the train—she has decided to come by bus. In this case, the "final cause"—the meeting at the train station—*never* exists. Yet it is still correct to say that the man went to the station *in order to* meet his wife. There is no contradiction in an unfulfilled purpose or an unrealized goal, but there is a contradiction in attempting to postulate in such cases the causal efficacy of non-existent and never-to-be-existent events.

As my analysis indicates, what causes and explains purposeful actions is not actually the future goal, but the agent's *present desire for* the future goal. The future goal does not exist yet—and may never exist—but the agent is able to conceive or imagine a future goal, and it is this present mental content which causes him to undertake the action he hopes will bring about his realization of that goal. In the preceding example, it is the husband's present desire to meet his wife plus his present (correct or incorrect) belief that by driving to the train station he will meet her that together cause his action and supply the basis for the action's explanation.

Thus in the case of purposeful action, the aspect of "final causation" presents no real problem. The action is caused and explained by one's present desire for some anticipated future consequence of one's action. This point has been made succinctly by R. B. Braithwaite:

> Teleological explanations of intentional goal-directed activities are always understood as reducible to causal explanations with intentions as causes; to use the Aristotelian terms, the idea of the "final cause" functions as the "efficient cause."[3]

An important consequence of this analysis is that "final causation" in the paradigm case of purposeful action is seen to be a species of ordinary "efficient" causation, rather than constituting an alternative kind of causation. Both philosophers and biologists have frequently assumed final causation and efficient causation to be irreducibly separate phenomena. It is often maintained that explanation via final cause and explanation via efficient cause are answers to two entirely separate questions. According to that view, the question "Why did this process occur?" may mean either "What is the efficient cause of this process?" or "What is the final cause of this process?" But in the case of purposeful action—the kind of case on which our very idea of final causation is based—these are not two *separate* questions. In purposeful action, the final cause (*qua* content of consciousness) *is* the efficient cause.

For instance, the question "Why did the man drive to the train station?" cannot be analyzed into two separate, irreducible questions, one seeking the final cause and one the efficient cause; both questions concern the efficient cause of the man's action. The explanation in terms of final causation would be: the (final) cause of his action was the meeting with his wife. The explanation in terms of efficient causation would be: the (efficient) cause of his action was his desire to meet his wife (plus his belief that his action could bring about the meeting). Obviously the two answers are the same except that in the latter we make explicit reference to the agent's mental states, whereas in the former explanation we take the agent's mental states for granted and cite only the existential objects of those states.

The final-cause explanation and efficient-cause explanation refer to one and the same causal chain. The difference in this context is only one of specificity: in the case of final-cause explanations of purposeful actions we are less concerned with the specific links in the causal chain than in the overall connection made by the chain as a whole. In the foregoing example, in saying the man acted as he did *in order to* meet his wife, we are certainly not denying the role of his beliefs and desires, nor are we attributing causal efficacy to possible future events—we are simply naming the object the desire for which caused the action.

Likewise, on the vegetative level, teleological explanation, I will argue, is not an irreducibly separate kind of explanation, but is rather a less detailed form of ordinary mechanical explanation in terms of efficient causes. (This position, however, does not reduce teleological explanation to the status of a dispensable abbreviation, any more than the recognition that animals are a sub-category of living organisms automatically reduces the concept "animal" to a dispensable abbreviation.) The problem of teleology concerns whether or not there is a distinguishable sub-category that may properly be termed "teleological" within the wider category of mechanical processes, and, if so, whether or not vegetative actions are teleological.

IV
SELF - GENERATION

Having developed an analysis of purposeful action, the next major section of this book will be devoted to an investigation of the similarities between purposeful action and vegetative action. The thesis to be defended is that each aspect of purposeful action that renders it teleological is also present (in a non-conscious form) in vegetative action.

Before comparing vegetative and purposeful action, it will be helpful to present some clear examples of vegetative processes to bring the comparison into focus. I will present one example from each of the three levels of vegetative action: processes occurring within the cell, gross processes performed by plants, and physiological processes occurring in animals.

CELLULAR RESPIRATION

A living organism does not have the passive stability of inanimate matter. In order to maintain its structure, the organism has to do work (in the thermodynamical sense). In order to do work, it, like any other entity, must have energy. Biochemist George Wald writes:

> How do present-day organisms manage to synthesize organic compounds against the forces of dissolution? They do so by a continuous expenditure of energy. Indeed, living organisms commonly do better than oppose the forces of dissolution: they grow in spite of them. They do so, however, only at enormous expense to their surroundings. They need a constant supply of material and energy to maintain themselves, and much more of both to grow and reproduce. A living organism is an intricate machine for performing exactly this function. When, for want of fuel or through some internal

40

failure in its mechanism, an organism stops actively synthesizing itself in opposition to the processes which continuously decompose it, it dies and rapidly disintegrates.[1]

The ultimate source of all the energy for all the living
actions on earth is the energy of sunlight. "Autotrophic"
organisms are those which have the ability to harness the
energy of sunlight directly, through photosynthesis, and to
store this energy in the chemical bonds of carbohydrates
(chiefly glucose) synthesized in the process. "Heterotrophic"
organisms are those which obtain their energy by eating the
nutrients synthesized by the autotrophic organisms (or by eating organisms who have themselves eaten autotrophic organisms).

Respiration is the process of extracting the energy stored in
the synthesized nutrients.

> Photosynthesis binds energy from solar radiation into
> complex organic compounds such as glucose and com-
> pounds synthesized from glucose. Before this potential
> energy can be utilized in the doing of work, either by the
> green plants or by some organism that has eaten the
> green plant, the large energy-rich molecules must be bro-
> ken down chemically and the energy released. This
> oxidative breakdown is called *respiration*; it is a process
> that must occur in every living thing, and is thus a funda-
> mental characteristic of life.[2]

Biologists have identified two basic types of respiration. In
anaerobic respiration, glucose is broken down into smaller
molecules in the absence of oxygen (thus "anaerobic"—"without air"). The process by which this is done is called glycolysis; glycolysis is performed by some types of bacteria and para-
sitic worms. In addition to glycolysis, the more common and
potent form of respiration involves the use of oxygen: aerobic
respiration. Aerobic respiration is more potent in the sense
of making available a greater percentage of the potential
energy stored in the carbohydrates.

Respiration is often compared to inanimate combustion,
as in the release of energy in an automobile engine by the

burning of gasoline. Actually, these two kinds of oxidation differ significantly. The first difference lies in the fact that in the case of respiration the fuel is produced by the organism itself. The organism takes in materials and/or energy from the outside environment and then processes them until they reach the proper chemical condition for respiration to take place. (In autotrophic organisms sunlight provides the energy to synthesize glucose from carbon dioxide and water; in heterotrophic organisms the raw food which is ingested must be prepared for respiration through the digestive process.) A closer mechanical analogy to respiration would be an engine which was equipped to drill for its own crude oil, and then to refine it to gasoline which it could then ignite.

But important differences between respiration and combustion remain: the oxidative process in respiration is enormously more complex than the simple one-step reaction involved in combustion. The simplest known form of respiration consists of at least eleven separate steps, each of which must be performed in proper sequence under narrowly controlled conditions, catalyzed by a specific enzyme.

> The oxidation of glucose in the [animal] cell proceeds in two major phases. The first, or preparatory, phase called glycolysis, brings about the splitting of the six-carbon glucose molecule into two three-carbon molecules of lactic acid. This seemingly simple process occurs not in one step but in at least 11 steps, each catalyzed by a specific enzyme. If the complexity of this operation seems to contradict the Newtonian maxim *Natura enim simplex est,* then it must be borne in mind that the function of the reaction is to extract chemical energy from the glucose molecule and not merely to split it in two.[3]

Another biologist contrasts respiration and ordinary combustion:

> A living organism is a machine that requires vast amounts of energy for the chemical and physical work it must perform. Organisms obtain their energy by the oxidation of biological materials—by burning food as fuel. The burn-

ing does not proceed in the random and inefficient way it would in a furnace; rather, enzymes act as catalysts and the oxidations take place in a controlled sequence of small steps in which more nearly the maximum obtainable energy is liberated.[4]

It is worthwhile pausing to note that even so "mechanistic" a process as cellular respiration gives the *appearance* of teleological causation. The biologists quoted describe the process in teleological terms; they speak of the "utilization" of energy, of the "function" for which the oxidation occurs, and frequently liken the organism to a machine—which means likening it to an object consciously designed so as to provide a desired result. In contrast, a physicist would not describe inanimate processes in such terms: he would not liken a rain cloud to a factory for the production of water droplets, or speak of a river system as "utilizing" the downward slope of the land in order to get water from the interior to the sea. Yet respiration is clearly an automatic, non-conscious, biochemical process. Each step of the process presently understood consists of ordinary chemical reactions which can be reproduced in the test tube and which are governed by the same laws which obtain in the case of inanimate chemistry. This is consistent with the emergentist view that teleological causation is an advanced form of mechanical causation, not an *alternative* mode of action. As stated in chapter 2, in my view the contrast is not between teleological causation and mechanical causation, but rather between simple mechanical causation and teleological mechanical causation.

PLANT TROPISMS

Compared to animals, plants seem to lead a static existence. But it is clear from our knowledge of cellular metabolism that plants are engaged in an ongoing process of intricately coordinated actions. In addition to such intracellular actions, plants have a definite range of gross behavior, such as those actions involved in their development from seed to maturity and in reproduction. Some of the most interesting and best understood of these gross actions are tropisms. Tropism is simply an orientation reaction—a bending or turning of the plant in response to some external stimulus (with no implication that the "stimulus" is perceived by the plants).

Perhaps the best known plant tropism is positive phototropism. This tropism consists of the plant's tendency to orient its leaves in a direction perpendicular to incident light. The result (or goal) of this tropism is to increase the plant's absorption of the light energy required for photosynthesis.

The mechanism underlying phototropism is fairly well understood; it involves the influence of a hormone (auxin) on the plant's growth rate.

> The explanation of the phototropic response is fairly obvious. Light must somehow affect the release of auxin by the tip. When the light strikes the plant from one side, it must reduce the auxin supply on that side. Consequently, the illuminated side of the plant grows more slowly than the shaded side, and this asymmetrical growth produces bending toward the slower-growing illuminated side.[5]

The same kind of hormone is also responsible for another kind of plant tropism—the tendency of plants to orient themselves in relation to the force of gravity—the shoots growing upwards and the roots growing downwards. The shoot of the typical plant exhibits negative geotropism: it tends to grow upwards against the force of gravity; the roots exhibit positive geotropism, growing downwards in response to gravity.

If such a plant is turned on its side, its direction of growth will compensate: the shoot will start to grow in the new upwards direction and the roots will start to grow in the new downwards direction.

Both positive and negative geotropism are caused by the fact that the force of gravity makes the growth-controlling auxins settle in the lower part of the horizontal stem or root with a resulting asymmetrical growth rate (figure 3).

gravity 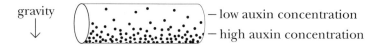 — low auxin concentration

— high auxin concentration

Surprisingly, the same asymmetrical auxin distribution has exactly opposite effects in shoot and root. In the shoot, the higher auxin concentration accelerates growth; in the root, the higher auxin concentration retards growth. This paradoxical effect is explained by the fact that root cells and shoot cells respond differently to the same concentrations of auxins; what is a growth-retarding concentration for roots is a growth-accelerating concentration for shoots. All cells are accelerated in their growth rate by a small amount of auxin, but above a certain concentration, each incremental increase in concentration produces a diminished effect, until a point is reached at which the addition of still more auxin actually retards the growth rate. The difference between the positive geotropism of the roots and the negative one of the shoots is explained by the fact that the growth-retarding concentration is reached far sooner for root cells than for shoot cells. A concentration of one part per million, for example, is strongly inhibitory of root growth while shoot growth is accelerated until concentrations of about one hundred parts per million are reached.

Here again we find a rather complicated, but *fully mechanical* process behind our pre-scientific notion that plants turn "in order to" obtain sunlight, that their roots "seek to reach" the deep earth where there are nutrients, and that their shoots grow upwards "so that" they can receive more sunlight. Any teleological account of plant tropisms will have to be compatible with this mechanical explanation.

THE HEARTBEAT

The circulation of the blood in the bodies of vertebrates is in general outline familiar to everyone today and does not require description. Our concern will be with the beating of the heart as the source of this circulation. Although the heartbeat is an action performed by animals which are conscious, it is not a conscious action and cannot be considered purposeful by our analysis of purposefulness.

Some of the physiological results of blood circulation by heartbeat are listed by a prominent cardiologist:

> The blood bathes the tissues with fluid and preserves their slight alkalinity; it supplies them with food and oxygen; it conveys the building stones for their growth and repair; it distributes the heat generated by the cells and equalizes body temperature; it carries hormones that stimulate and coordinate the activities of various organs; it conveys antibodies and cells that fight infections.[6]

I will argue that besides being the effects or consequences of the heartbeat, the foregoing represent the functions which constitute the heartbeat's *raison d'être*: the heart beats in order to achieve these results. Mechanists hold that since the heartbeat is a vegetative action this teleological explanation is nonsense, and we must restrict ourselves to the language of cause and effect rather than that of means and ends.

SELF - GENERATED ACTION

In making a distinction between "action" and "process" earlier, I touched on the first similarity between vegetative and purposeful actions: both are "self-generated."[7] The self-generated nature of all living processes—whether conscious or vegetative—is one of the most obvious characteristics of life. Reference to this characteristic can be found at least as early as Plato, in the *Laws*: "when the thing moves itself, we speak of it as *alive*."[8] In *De Anima*, Aristotle gives a clearer statement of the same idea, holding that for an entity to have life (*psyche*) it must be one "having *in itself* the power of setting itself in action [*kinesis*] and arresting itself" (translator's emphasis).[9] In the *Physics*, he says of non-living objects: "It is impossible to say that their action [*kinesis*] is derived from themselves: this is a characteristic of life and peculiar to living things."[10]

The idea of self-generated action is inherent in the concept of goal-directed, teleological causation. An entity which is incapable of moving itself, an entity which can respond only passively to the pushes and pulls of external forces, cannot be conceived as *directing its action*. A falling stone is pulled toward the earth by gravity, but in such a case the stone is not *acting* at all, much less *directing* its motion toward its final state.

Returning to the teleological paradigm of man's purposeful actions, we can readily see that self-generation is a necessary condition for purposefulness. A purposeful action is one which a *man performs*, not just any motion his body undergoes. In support of this point, consider a case in which all the elements of purposefulness except self-generation are present. A man dives off a cliff into a lake. The whole action, under the description "diving into the lake," is purposeful: it is an action which the man performs because of his desire for its anticipated consequences. Now, however, consider only a small portion of the dive: a segment of, say, one-half second during the middle of his descent. We have already seen (*supra*, chapter 1) that purposeful actions can involve sub-processes which are not purposeful (e.g., walking is purposeful, but the contraction of individual muscle fibers in the leg is not). Similarly, the half-second segment of the dive is not purposeful because it is not self-generated. Once the man is in the air, the overall motion of his body is determined solely by external forces such as gravity and air resistance. Although his descent in that half-second was caused by his desire for its anticipated consequences, that segment of the action *per se* is not purposeful (nor is it vegetative; it is mechanical in the same sense that a stone's fall is).

To make the contrast fully clear, consider, during this same half-second segment, not the diver's gross downward motion, but the relative motion of the parts of his body as he executes the maneuver required for a given kind of dive (e.g., a jack-knife dive). Although once in mid-air he cannot control the gross downwards motion of his body as a whole (nothing he does can noticeably alter the path of his center of mass), he does control the relative motions of the parts of his body, and these relative motions, being self-generated actions, are purposeful. Given the choice to make the dive, the fall from point A to point B in the half-second interval is not self-generated and not purposeful, but his use of his limbs to execute a certain maneuver during the fall from A to B is self-generated and *is* purposeful.

Granted that self-generated action is a necessary feature of teleological causation, we must now analyze what self-generated action consists of and must show that all vegetative actions satisfy this analysis.

In order to go beyond Aristotle's characterization of self-generated actions, we must take into account several rather sophisticated factors. Given our modern understanding of physical processes, what does it mean to say that living organisms have in *themselves* the power to set themselves in action?

The main point is this: in self-generated actions the *energy* for the action is supplied by the organism itself from a source within and integral to its own body. In a non-self-generated action the energy for the action is supplied by some external force. For instance, when a man moves his arm, the energy for this action is supplied by the man's muscles, whereas in the case of a dead leaf that is blown across the ground, the energy is supplied entirely by an external factor: the wind.

Ultimately, of course, energy cannot be *created* by an organism: all of a living organism's energy resources are derived from what it takes in from the environment. The essential criterion, then, for self-generation is that the *immediate* source of the energy for the action be located within the organism, not externally. It is the fact that living organisms possess a self-contained store of energy which manifests itself in the simple observation that they move themselves instead of being merely pushed, pulled, and shoved about by external forces in the manner of inanimate objects.

Secondly, if the action is to be self-generated, the energy source must belong to the organism as a whole and only secondarily or derivatively be associated with any given part or any given process. With self-generated action there is a certain crucial independence of energy supply and action; the same energy supply is available to drive any one of a number of different actions. (In addition, the same action can often be fueled by different energy sources.) The energy supply belongs to the organism as a whole, as manifested by the fact that the organism can utilize the same energy supply in a number of alternative ways. An animal's energy supply can be utilized in muscle contraction, or in self-repair, or ultimately even for energy-requiring processes in other parts of the body.[11]

Whenever this independence of energy supply and action exists, there must be some means of controlling the utilization of the energy. That is, if in self-generated action the same energy supply is available to power alternative processes, then the organism must have some mechanism or faculty which channels the energy in the required direction. The idea of such a *directive mechanism* is implicit in the idea of self-generated action. In man, for instance, it is the nervous system (and/or consciousness) that is the directive mechanism which controls the operation of all the muscles and many of the glands of man's body. Or, on the cellular level, many biochemical directive mechanisms exist in a hierarchy terminating in the DNA molecule. It is the existence of such directive mechanisms that provides the separation of "self" and action which enables us to conceive of an organism as acting self-generatedly—i.e., as "having *in itself* the power of setting itself in action and arresting itself." (Again, my use of the term "directive mechanism" is not meant to imply either the presence or the absence of consciousness or volition; nor do I mean by using the adjective "directive" to be bringing in a teleological connotation at this stage.)

This idea of a directive mechanism that mediates between an energy source and alternative possible actions allows us to distinguish self-generated actions from other processes characterized by a release of internally stored energy. For instance, consider the expansion resulting from the release of a compressed coil spring. It is true that the release of the potential energy stored in the spring supplies the energy for its subsequent expansion to its equilibrium length, but the potential energy of the spring is available for only that one action—the energy in this case is associated primarily with the action rather than with the "self" (i.e., the spring). There are no directive mechanisms mediating between the energy supply and the expansion; one could not speak of the spring as being "programmed" to expand (whereas one could speak of a plant as being "programmed" to turn its leaves toward the sun).[12] The same is true of chemical reactions or atomic reactions which are energy-releasing on net balance. Such reactions are not self-generated because their energy output cannot be channeled by a directive mechanism to power alternative kinds of processes.

In a provocative article, "Living and Lifeless Machines," R. O. Kapp makes another observation concerning the manner in which a living organism's energy supplies are associated primarily with the whole organism and only secondarily with any given kind of action. Having made the general point that a living organism in contrast to a man-made machine, utilizes one and the same components for radically different engineering functions (i.e., as moving parts, as braces, and as energy sources), he says the following of the energy source in particular:

> Let us now turn our attention to the fuel. In a motor car this is petrol and it is stored in a tank. In the human body it is chiefly glycogen and is stored in the muscles, having been converted from glucose in the liver. The chemical processes during muscle activity include the combination of muscle protein with sodium. This protein is therefore another part of the fuel. So both the glycogen and the protein serve the double function of being fuel and being constituents of those muscle fibres that are at one moment moving parts and at another components of the frame. The living body is analogous to a motor car in which the chassis, brakes, cylinders, pistons, connecting rods, valves, and bearings all contained combustible material, some of which was burnt whenever the driver placed his foot on the accelerator.[13]

Here, as earlier, it must be noted that the independence of energy source and action does not imply a "transcendence" of physical law. The fact that the same energy source is available for different actions does not imply that the organism possesses an ability to disobey the thermodynamical laws obtaining in the case of inanimate objects: the channeling of energy by the directive mechanism will require that the sub-processes follow the path of least resistance (in thermodynamical terms, toward states of lower "free energy").[14]

The issue here, as in every other case of "emergent properties," is simply one of complexity of organization. If a self-generated action is analyzed deeply enough, it will be found to consist of an orchestrated sequence of biochemical reactions which are themselves energy-releasing on net balance in the same way as for inanimate chemical reactions that are exothermic. The self-generated aspect comes about from the highly ordered integration of these reactions.

For our purposes, it is not necessary to go into an extensive analysis of the details of the physical principles underlying the distinctive nature of self-generated action. My purpose in mentioning some of the technical issues involved has been to indicate that the common-sense idea that organisms move themselves is not illusory, but rather has a sound foundation in physical fact. Although I believe it is possible to defend the position that self-generated action is possible only to living organisms, it is not necessary to do so. The relevant point is not that inanimate processes are without exception non-self-generated, but that both vegetative and purposeful actions *are* self-generated. The fact that some non-living objects may exhibit some of the features of living action is beside the point. Being self-generated is the first similarity which unites vegetative and purposeful action. The existence of borderline cases, the difficulty of classifying the actions of certain hypothetical machines, or the existence of exceptions to the general rule, would not invalidate the easily perceivable difference between those actions which are internally generated and those which are generated by an external factor. It is quite clear that all living action, whether conscious or vegetative, is self-generated in this sense.

Before applying this analysis of self-generated action to vegetative processes, it is necessary to consider an objection to the idea that purposeful actions are self-generated. It might be thought that purposeful actions by men and the higher animals are not actually self-generated, since they are caused by the perception of some external stimulus, and every such perception involves the reception of energy from the external environment by a sense organ. For instance, if a man reaches for a stone, this action is made possible by his perception of the stone, which is in turn made possible by the energy of the light which strikes the retina of his eye.

This objection, however, confuses self-generated action with *volitional* action. By "self-generated" no involvement of free will is assumed. In the preceding example, it is true that the stimulus energy *triggers* the man's reaching action, but the stimulus is not the source of the energy for the performance of the action itself. This is clear from the fact that there is no necessary relation between the magnitude of the stimulus energy and the magnitude of the energy expended in the sub-sequent action. A man may put forth the same amount of energy in reaching for the stone when he sees it in the sun-light as he does when it is only faintly illuminated. The ener-gy of the external stimulus can vary by a tremendous factor without in the least affecting the level of energy expended in the subsequent action. In contrast, the speed with which a dead leaf is blown across the ground is a direct function of the physical pressure of the wind which pushes it; the leaf's motion cannot be said to be *triggered* by the wind.

(The phenomenon of "triggering," as I am calling it, is wider than that of self-generated action. Many inanimate actions involve the release of potential energy in a non-self-generated fashion in reaction to a relatively tiny amount of triggering energy supplied from the outside. The firing of a gun clearly exemplifies a situation in which the force exerted upon the trigger bears no necessary relation to the velocity of the bullet upon firing. But although triggered, the gun's fir-ing is not self-generated because the energy source is not inte-gral to the acting entity as a whole nor is the energy source available for any other kind of action.)

SUMMARY

It is evident that each of the three examples of vegetative processes qualifies as self-generated. The heartbeat is the rhythmic contraction of a muscle which, like all other muscular contractions, is accomplished by the utilization of energy stored chemically in the body (i.e., the combination of oxygen carried to the heart in the bloodstream with glycogen stored in the cardiac muscle itself). The heartbeat is an active process under the control of many directive mechanisms (e.g., the "pacemaker").

Plant tropisms consist of asymmetrical growth rates under the immediate control of the auxin mechanism already discussed. Thus these tropisms are as self-generated as the process of growth itself.

Cellular respiration, the oxidative breakdown of the organism's fuel, is self-generated in two respects. First, the net effect of respiration is to use the energy stored in a starch (e.g., glucose) to add an extra phosphate group to adenosine diphosphate (ADP), thus converting it to adenosine triphosphate (ATP). The starch constitutes one of the organism's energy "savings accounts" which must be converted into the "cash"—the ATP—which is expended in doing actual biochemical work.

Hence respiration is self-generated in that its product (ATP) is synthesized through the release of internally stored energy under the control of directive mechanisms (e.g., the mitochondrion). But in addition it is self-generated in respect of the fact that some ATP must be "spent" (converted into ADP) in the very process of creating more ATP. In respiration ATP is used to *drive* the process by which more ATP is formed. Respiration is *not* a simple release of energy as in the expansion of a coiled spring; rather, it contains some steps in which the process must be "pushed uphill" by the energy-releasing conversion of some ATP to ADP.

To summarize: for millennia, men have made the distinction between cases in which an object moves itself and cases in which an object is set in motion by some outside object or force. At an advanced level of knowledge, we recognize that every physical process requires a source of energy, and hence can reformulate this distinction in terms of the presence or absence of an internal energy source. Accordingly, we observe that living organisms characteristically do, and inanimate objects characteristically do not, have the power to generate their own actions. Recognizing that there are inanimate processes which result from the release of an internal store of energy, but which are still importantly different from living actions, we state another property of self-generated actions: the energy must be integral to the self, the agent; the agent's energy resources must be prior to any one kind of action; the agent itself must possess the energy which it can then use to power any of a number of different kinds of actions, according to the functioning of a directive mechanism(s).

Thus, at the more advanced level of knowledge, we can say that an action is *self-generated* when it results from the utilization of an internal energy source integral to the agent according to a directive mechanism. Self-generated action, thus understood, constitutes a fundamental similarity between vegetative and purposeful actions. This is the first similarity uniting vegetative and purposeful actions with regard to mode of causation. By itself, however, the fact that vegetative actions, like purposeful actions, are self-generated is not sufficient to show that any vegetative actions can be classified with purposeful actions as teleological. Two more such similarities remain to be discussed in the next three chapters.

V
VALUE-SIGNIFICANCE

THE APPARENT PURPOSEFULNESS OF VEGETATIVE ACTION

The problem of teleology arises because vegetative actions appear to be similar to purposeful actions. This appearance is so forceful that many imagine that plants are conscious, experiencing pleasure and pain, and that they will flourish only when given affection. Or, consider the basis of the appeal of the Argument from Design. Vegetative processes such as the embryonic development of the human eye, the functioning of internal organs, and the healing of bodily wounds were almost universally ascribed, prior to Darwin, to the intent of a Designer.

What misinterpreted facts lie behind this impression? What properties of vegetative action lend it its apparent purposefulness?

We have already noted one such property: vegetative actions have in common with purposeful actions the property of being self-generated. But self-generation is only a necessary condition for apparent purposefulness. If a concept of "goal-directed action" wide enough to embrace both vegetative and purposeful action is to be defined, "self-generated action" will supply only the genus; "self-generated action" denotes a wider class than does "goal-directed action." There are many cases of self-generated action which are not teleological—i.e., not goal-directed. For example:

1. The production of oxygen by green plants in photosynthesis.
2. The heart's production of a thumping sound.
3. The extension of the lower leg in the human patellar reflex.
4. The growth of a tumor.

Examples 1 and 2 cite incidental effects of self-generated actions, not the goals for the sake of which those actions would normally be said to occur. Examples 3 and 4 cite self-generated actions which have no known goal at all. These examples suffice to show that self-generated action is not to be equated with goal-directed action.

What feature, besides self-generation, could account for the apparent purposefulness of vegetative actions?

"PLASTICITY"

Some writers have sought the answer to this question in terms of the persistence with which the same end state is reached by the organism despite varying initial conditions and obstacles placed in the path of goal-attainment. R. B. Braithwaite has emphasized the teleological significance of this "plasticity" of response:

> It seems impossible to find any characteristic of the final state by itself of a teleological causal chain which is general enough to cover all the goals of goal-directed action and yet specific enough to differentiate such actions from other repeated cycles of behavior. It is necessary, I think, to look at the whole causal chain and not merely at its final state.
>
> It seems to me that a distinguishing criterion can be found in one of the characteristics which biologists have emphasized in their descriptions of goal-directed behavior, namely persistence towards the goal under varying conditions. To quote E. S. Russell: "Coming to a definite end or terminus is not per se distinctive of directive activity, for inorganic processes also move towards a natural terminus. . . . What *is* distinctive is the active persistence of directive activity towards its goal, the use of alternative means towards the same end, the achievement of results in the face of difficulties." [*The Directiveness of Organic Activities*, p. 144]. . . Plasticity is not in general a property of one teleological causal chain alone: it is a property of the organism with respect to a certain goal, namely that the organism can attain the same goal under different circumstances by alternative forms of activity making use

frequently of different causal chains.[1]

Although "plasticity" is indeed a characteristic feature uniting vegetative and purposeful actions, it is a relatively superficial one. "Plasticity" is not the characteristic responsible for the apparent purposefulness of vegetative actions, nor does "plasticity" *constitute* goal-directedness. Goal-directed actions exhibit "plasticity" *because*, as I will subsequently argue, they are caused and controlled by the goal rather than the initial conditions of the set-up.

The plasticity criterion clearly includes too wide a class of processes as teleological. For instance, the heart's production of a thumping sound is a "final state" achieved with the same degree of plasticity as the beat's apparent goal (or function): circulation of the blood. In the process of respiration, cells exhibit exactly as much plasticity with regard to the production of water and carbon dioxide as they do in the production of ATP, yet only the latter is ever described as the goal or function of respiration.

A *persistent result*, although characteristic of goal-directed action, is not equivalent to a *goal*. (For a further discussion of the inadequacies of the plasticity definition, see Israel Scheffler, *The Anatomy of Inquiry.*[2])

In explaining the apparent purposefulness of vegetative actions, let us look again at our analysis of the properties which constitute purposefulness: A *purposeful* action is a self-generated action caused by the agent's desire for some anticipated consequence of his action. (In the light of the points made in the preceding chapter, I have added explicit reference to self-generation to the analysis given in chapter 3.)

Now consider the two effects of the heartbeat: circulation of the blood and production of the thumping sound. What differences between these two effects explain why only blood circulation is anthropomorphically viewed as the "purpose" of the heartbeat? Let us look at it this way: if the heartbeat had to be performed as a conscious, intentional action by man, rather than occurring as an automatic vegetative action, which effect would be the purpose for the sake of which man would beat his heart—to circulate his blood, or to produce a thumping sound? Clearly we would expect a man to beat his heart in order to circulate his blood. We would automatically assume that a man would *desire* to have his blood circulate, since this is necessary to his survival; we assume his indifference toward the thumping sound that emanates from his chest.

NEEDS, BENEFITS AND VALUE SIGNIFICANCE

Even though the beating of the heart is in fact a vegetative process, rather than a purposeful action caused by a desire for some anticipated end, we have identified one of the major reasons why the heart's circulation of the blood *resembles* purposeful action: the circulation of the blood is beneficial to the organism.

In vegetative action, though the absence of consciousness precludes the existence of emotions and evaluations, there is nonetheless something that plays a role similar to that played by desires in the case of purposeful actions: *needs*.

Although the (apparent) goals of vegetative actions are not *desired* by the organism, they are *needed*. It is partly (though not exclusively) the fact that blood circulation is beneficial to the organism—i.e., that it serves a need of the organism—that explains why we are tempted to view blood circulation as if it were the purpose of the heart's action.[3]

"Need," in the sense here employed, is one of a family of terms that pertain to the beneficial relationship of things to the organism. Other terms in the same family include: "useful," "benefit," "value," "utility," and "advantage." For instance, rather than saying the circulation of the blood is needed by the organism, one could say that blood circulation is beneficial to the organism, useful to the organism, has utility for the organism, or is advantageous to the organism. Although they are not synonyms, each of these terms evaluates something in terms of its effect upon the acting organism. To put it in a general way, these terms describe the *value-significance* of things for the organism.

In purposeful action there is also a value-significance placed upon the end state: the end state is desired (or consciously valued). In desiring something, a man (or an animal) places a psychological value-significance upon the outcome of the attempt to achieve the desired object. Successful achievement of the desired object is associated with pleasure, happiness, satisfaction, etc.; failure is associated with pain, suffering, frustration.

The fact that the ends of both vegetative and purposeful action have a value-significance for the agent constitutes the second general similarity between the two types of action, and sets them off as a class from both the processes of inanimate matter and the non-teleological actions of organisms (e.g., the heart's production of the thumping sound).

"TO WHOM?" AND "FOR WHAT?"

In order to see this more clearly, let us investigate the basis of the concept of "value-significance." In a similar connection, Ayn Rand has observed:

> The concept "value" is not a primary; it presupposes an answer to the question: of value to *whom?* and for *what?* It presupposes an entity capable of acting to achieve a goal in the face of an alternative. Where no alternatives exist, no goals and no values are possible.[4]

Let us look at each of these points separately.[5] First, in purposeful action to whom is the value of the goal a value? Whose value-significance are we concerned with in judging an action to be purposeful? Clearly we are judging by the action's value-significance for the agent, rather than by the action's value-significance for other individuals. For example, assume Smith desires to kill Thompson, and consequently pushes him off a cliff. Thompson's fall is not a purposeful action of Thompson's—it does not result from his desire to die (he has no such desire). Thompson's fall is a purposeful action of *Smith's*, since it was caused by Smith's desire for the death of Thompson, which he expected to be the act's result.

The situation is, in this regard, the same as if Smith were pushing not Thompson but a boulder off the cliff. The value-significance of the action for Thompson is relevant only insofar as it is in turn a value-significance for Smith, the agent.

The case is the same in altruistic actions. If a man gives a dollar to a beggar, the action is purposeful by virtue of the man's desire to help the beggar, not simply by virtue of the fact that the beggar is helped. Suppose the man does not desire to help the beggar; suppose he despises the beggar and wants to see him suffer, but gives him the dollar anyway in order to impress a priest who is viewing the scene. In this case, the actual purpose of the man's action is not the help given to the beggar, but the favorable impression he hopes to make upon the priest. Despite the positive value-significance of the man's action for the beggar, the help rendered the beggar is merely a distasteful consequence grudgingly accepted in order to achieve the man's actual purpose: impressing the priest. We may even imagine that after the priest has left, the man demands that the beggar return the money.

The satisfaction or frustration that counts in making the action purposeful is that of the agent; the value-significance required is that of the agent. Accordingly, the value-significance which imparts the apparent purposefulness to vegetative action is that of the acting organism. The heartbeat of an animal is beneficial to that animal. The tropism of the plant is beneficial to that plant. The process of respiration performed in the cell is beneficial to that cell. Frequently, such actions also involve benefits to other organisms besides the agent, but those external benefits are not what lend the actions their apparent purposefulness.

This leads us to the second question which must be answered in judging the value-significance of an action: for *what?* In terms of what is the benefit to the agent judged to be a benefit? In terms of what requirement is the end state needed? What is *at stake* in the action in terms of which success or failure is judged to be success or failure?

In the case of purposeful action, the stake is on the first level a conscious one: pleasure, satisfaction, or, in the case of man, the achievement of something one considers good to have. But on the vegetative level, pleasure, satisfaction, and judgments of value do not enter. What then is at stake on the vegetative level? In terms of what effect upon the acting organism do we say vegetative actions are beneficial, needed, advantageous, useful, etc.?

To put the question in a concrete case: why do we judge blood circulation to be beneficial to the organism? We might answer: because blood circulation supplies the body's cells with nutrients and oxygen. This merely raises the further question: why do we judge the cells' possession of oxygen and nutrients to be beneficial to the organism? We can answer: because it makes possible the metabolic processes of the cells. Again the question is only deferred: why do we judge metabolism to be beneficial to the organism? Here the answer is: because it aids the organism's life—the organism must carry out the metabolic processes in order to remain in existence. Beyond this we cannot go—it makes no sense to ask: why is maintaining the organism's existence beneficial to the organism?

THE ALTERNATIVE OF LIFE VS. DEATH

What we were seeking in the series of questions was an answer to the basic question: what value does blood circulation have for an animal—why does it *need* blood circulation? In tying blood circulation to the very existence of the organism we have answered that question, for "to be beneficial to an organism" and "to be beneficial to an organism's life" mean the same thing. Similarly, "to be needed by an organism" and "to be required for the organism's existence" mean the same thing. Thus, as this example illustrates, *all value-significance on the vegetative level is relative to the organism's life.* In this context, "value-significance" means "survival-significance." The organism's life is the implicit ultimate value by reference to which the value-significance of all other states is determined. As George G. Simpson states, "The over-all and universal goal is a posteriori at the given moment and is simply survival, which involves comparative success in reproduction."[6]

Hence a necessary (though not, as we shall see, sufficient) condition for the end state of an action to qualify as the action's *goal*, rather than merely as its effect, is that it be needed by the organism for survival. (This does not mean that being deprived of the goal must spell instant death, only that being thus deprived increases its chances of dying.)

Need is present only when an entity faces an alternative. If there is no alternative faced in the action, if the organism's existence is neither helped nor threatened by the outcome of the action, the action cannot be said to be needed by the organism.

For example, if the organism's existence were not affected by the issue of whether or not its blood circulates, we could not say blood circulation is needed by or beneficial to the organism, and we could not ascribe any value-significance to blood circulation as the effect of the heartbeat. In this case blood circulation and the production of heart sounds would be on a par: neither would be seen as having an apparent purposefulness.

The chain of reasoning may be summarized as follows: apparent purposefulness is based on value-significance; value-significance is based on need; need is based on the alternative of the organism's existence or non-existence. Only living organisms face an alternative in terms of which self-generated action is needed. What is distinctive about living organisms is that their existence is conditional upon their successful performance of specific courses of self-generated action. As Rand states:

> Where there are no alternatives, no values are possible. There is only one fundamental alternative in the universe: existence or non-existence—and it pertains to a single class of entities: to living organisms. The existence of inanimate matter is unconditional, the existence of life is not: it depends on a specific course of action. Matter is indestructible, it changes its forms, but it cannot cease to exist. It is only a living organism that faces a

constant alternative: the issue of life or death. Life is a process of self-sustaining and self-generated action. If an organism fails in that action, it dies; its chemical elements remain, but its life goes out of existence. It is only the concept of "Life" that makes the concept of "Value" possible.[7]

The existence of a living organism is conditional—conditional upon its ability to satisfy its needs through its own self-generated actions. The existence of inanimate objects is not conditional upon their actions: (1) inanimate objects are not capable of self-generated action, and (2) they will continue to exist as long as they are not acted upon by external forces. A living organism faces the constant alternative of life or death—not simply in that it can be annihilated by an external catastrophe (as, for instance a stone can be pulverized by an advancing glacier)—but in that unless it can utilize the materials and energy in its environment to fuel the complex internal processes of self-maintenance, it will disintegrate. Biochemist Albert Lehninger writes:

> A living cell is inherently an unstable and improbable organization; it maintains the beautifully complex and specific orderliness of its fragile structure only by the constant use of energy. When the supply of energy is cut off, the complex structure of the cell tends to degrade to a random and disorganized state.[8]

The factor which sets the phenomenon of life apart from the inanimate world is the need of living organisms to act to obtain materials and energy from the environment and to utilize these items for self-maintenance.[9] Biologist Walter Bock has suggested that living organisms exhibit at least three distinguishing features:[10]

> 1. living organisms take in materials and energy from the environment;
> 2. they use the appropriated materials and energy for self-maintenance, self-repair, and self-reproduction;
> 3. once they have died, they cannot be reconstitut-

ed—failure is irreversible.

It is precisely this need of organisms for self-sustaining action that makes possible the distinction on the vegetative level between a goal and incidental effects; a goal is some condition sought for the sake of its ability to satisfy the survival needs of the acting organism.

The goal of any action should always be seen in the context of a hierarchy of goals, the value-significance of any given goal being derived from the fact that it serves as a means to (i.e., facilitates the attainment of) some further goal. For instance, the value-significance of sunlight as the goal of a plant's phototropism derives from the fact that sunlight serves as the means to the further goal of glucose synthesis. This hierarchy of goals implies the existence of an ultimate goal to which all the actions contribute. As we have observed, in vegetative action this ultimate goal is survival.

No single vegetative action is an end in itself nor is valuable in itself; all value-significance is relative to the survival of the organism as a whole, this being the ultimate end in relation to which all lesser goals are the means.

A common misconception is that of thinking of "survival" as if it were some single vital action that occurs after all the other actions have been completed. "Survival," however, means the continuation of the organism's life, and the organism's life is an integrated sum composed of all those specific actions which contribute to maintaining the organism in existence. In this sense in living action the parts are for the sake of the whole: the specific goal-directed actions are for the sake of the organism's capacity to repeat those actions in the future.

An ultimate goal, if it is truly ultimate, must be an "end in itself." An "end in itself" gives the appearance of a vicious circle: it is something sought for the sake of itself. This circularity vanishes when we regard *life* as an end in itself: actions at a given time benefit survival, which means they make possible the organism's repetition of those actions in the future, being then again directed toward survival, which means their further repetition, and so on.

All the goal-directed actions together as a hierarchical system constitute the life of the organism. For instance, the plant's tropism, being a self-sustaining action, is *part of* the plant's life, not merely a means to life. Thus life is, in the classic sense, an end in itself: it is sought for its own sake. *Life is the goal of life*: the continuation of life in the future is the goal of all the actions which constitute the organism's life in the present. Rand states:

> It is only an ultimate goal, an *end in itself*, that makes the existence of values possible. Metaphysically, *life* is the only phenomenon that is an end in itself: a value gained and kept by a constant process of action.[11]

As to the actual consequences of living action, viewed from an empirical standpoint, Theodosius Dobzhansky, one of the founders of modern evolutionary theory, writes:

> Almost everything which an organism does, physiologically and behaviorally, serves to enable this organism to stay alive and to resist destruction.[12]

Biologists view the actions of organisms as "adaptive," and in so doing they are implicitly or explicitly ascribing a value-significance to the outcomes of these actions, a value-significance based on survival. Dobzhansky notes that "The structures and functions of living bodies are said to exhibit adaptedness, or end-directedness, when they are shown to contribute to individual survival or to reproduction."[13] Simpson is even more explicit:

> An adaptation is a characteristic of an organism advantageous to it or to the conspecific group in which it lives. . . .
> It is obviously advantageous for an individual to remain alive and for a group to continue reproducing itself—or, if this does not seem obvious, we will define this as a pertinent meaning of "advantageous."[14]

Thus we appear to have an answer to the difficulty raised

by Braithwaite:

> It seems impossible to find any characteristic of the final
> state by itself of a teleological causal chain which is gen-
> eral enough to cover all the goals of goal-directed action
> and yet specific enough to differentiate such actions
> from other repeated cycles of behavior.[15]

Such a characteristic does exist: the final states of those
actions we regard as teleological have value-significance. In
purposeful actions this value-significance is experienced consciously—e.g., via pleasure and pain. In vegetative actions,
this value-significance is biological and is based on the alternative of life or death.

PSYCHOLOGICAL AND BIOLOGICAL VALUE SIGNIFICANCE

An issue of crucial importance to the general defense of
vegetative teleology is whether or not the term "value-significance" is being used univocally here—i.e., whether conscious-
level value-significance and biological value-significance are
merely analogous or have a common essential nature. I will
argue in chapter 8 that "psychological" (as I will call it) value-
significance is in a sense reducible to biological value-significance based on survival. For the present it will suffice to
make a few observations on the close relation between the two
phenomena.

Firstly, as a general rule, the things which do in fact bring
pleasure to a conscious organism are the things which benefit
its survival, and the things which do in fact bring pain to a
conscious organism are the things which harm or endanger
its survival. (The only organism for whom we have *direct*
knowledge concerning what is pleasurable or painful is man,
but clearly the following remarks are applicable to animals as
well.) For example, eating nutritious foods is generally pleasant; eating spoiled (and hence harmful) food is repugnant;
injury to the body causes pain; bending a limb of the body to
a degree approaching that which causes fracture brings pain;
deprivation of food, water or air causes pain (and the pain

increases as the danger to survival increases); replenishing the body with food, water, or air brings pleasure or relief.

The correlation between things bringing pleasure and things promoting survival is not absolute: arsenic, for example, tastes sweet. But lack of correlation is the very rare exception, not the rule. Furthermore human beings can actively seek goals which they know will bring more pain than pleasure, as in the case of action from duty, but this kind of motivation is not essential to our understanding of purposeful action—the other animals are strictly "hedonistic," but their actions are no less purposeful for it. (The special case of human sexual masochism clearly represents a departure from the normal, and the common explanation of masochistic behavior is that a limited pain is sought for the sake of a momentary release from a deeper pain or anxiety.)

Secondly, the achievement of pleasure requires that the animal be alive; a dead animal or man cannot experience pleasure. Where the biological value of life is not achieved, the possibility of psychological values vanishes. Human beings, knowing this fact, place a psychological value on their own survival and on the things they see to be required for their survival. Many of the things that men do are explainable in terms of their desire to live (or at least their desire to avoid death). Thus for man, since (a) things which promote his survival do tend to bring pleasure or satisfaction, and (b) he knows that even those pleasures which may be unrelated to his survival can only be gained and enjoyed if he is alive, anything with biological value-significance will tend to have psychological value-significance in turn.

Things with psychological value-significance are, in general, things with biological value-significance. The things both men and animals desire are generally the things mentally and/or causally associated with their survival. Furthermore, survival is a necessary condition for pleasure. These observations may serve to postpone, if not fully answer, the objection that the term "value-significance" is used equivocally in its application to vegetative and purposeful actions. A thorough answer to this objection will be given in chapter 8.

GOALS AND INANIMATE PROCESSES

I have argued that only an entity needing to act to pre-serve its existence can have goals. An important implication of this is that teleological concepts are applicable only to liv-ing organisms. Many contemporary writers have proposed definitions of teleological concepts that would include certain non-living processes as teleological. For instance, Rosen-blueth, Wiener, and Bigelow define "purposive behavior," by which they include both conscious and non-conscious pro-cesses, in terms of feedback mechanisms which are "directed to the attainment of a goal, i.e., to a final condition in which the behaving object reaches a definite correlation in time or in space with respect to another object or event."[16]

But the "final condition" thus described is not equivalent to a *goal*. As Richard Taylor has pointed out,[17] every process in nature involves a "behaving object" that "reaches a definite correlation in time or in space with respect to another object or event." *Every* process is the process of some object which ends up somewhere at sometime. To define "goal" in a way that makes every result of every process a goal, is to empty the concept of "goal" of its specific meaning. As a teleological concept, "goal" has its ultimate origins in the case of man's purposeful actions. Its original meaning derives from and depends upon the distinction between those "final condi-tions" of human actions which are purposefully sought and those which are only accidental or incidental results. To define "goal" cybernetically, as Rosenblueth et al. do, is to wipe out the distinction on which the concept of "goal" rests.

It might be objected by an advocate of the cybernetical definition that I have misinterpreted the logic of the pro-posed definition. This objection holds that the criterion is not whether the object reaches some final condition or other (since all behavior satisfies that condition), but whether the object reaches some single *antecedently specified* final condition, rather than any other. This interpretation, however, raises the question of the means by which the final condition that "counts" is to be specified (obviously, it would be circular to say that the relevant final condition is the one which is the behavior's goal).

There are two alternatives: (1) the target may be specified simply by reference to that end state of the behavior which, according to the laws of physics, can be predicted to occur. On this alternative, however, it remains true that every deterministic physical process is rendered teleological by the cybernetical definition. It would be true, for example, that rocks have the *goal* of reaching the ground when dropped from a height, since reaching the ground would be an antecedently predicted "final condition in which the behaving object reaches a definite correlation in space or in time with some other object [the ground]."[18] (2) Alternatively, we may specify the relevant final condition independently of our prediction concerning where the object will in fact end up, but in this case there is no objective basis for deciding which of the logically possible final conditions is the goal and which are not. The matter becomes arbitrary and subjective: two individuals observing the same process may stipulate different final conditions as the goal. Consequently, an outcome which is successful goal-attainment for one observer is goal-failure for the other, and vice versa. It is clear that if neither observer is objectively wrong, if any logically possible final condition of a process may be considered its goal or not according to the subjective preferences of an outside observer, then this means *the process itself* has no goal—that goals are a subjective idea anthropomorphically projected onto neutral goal-less reality. In this case, if no human beings existed to make the required projection, no final condition would count as a goal.

Thus on one interpretation of the cybernetical definition, all natural processes have goals (since they all reach a definite correlation with some object or other), and on the other interpretation no natural processes have goals (since an end state is or is not a goal only relative to an arbitrary human decision to consider it as such). Neither interpretation results in a concept of "goal" that does the job that concept is intended to do: on neither interpretation is it possible to distinguish between those final conditions which are goals and those which are not.

One essential element that is missing from the cybernetical analysis of the concept "goal" is *value-significance*. The final conditions of the behavior of feedback mechanisms, such as target-seeking torpedoes, are not *needed* by the mechanism. Attainment of the "goal" does not benefit it, nor does failure to attain the "goal" harm it. In fact, since a feedback mechanism faces no alternative, the concepts expressing value-significance, such as "benefit" and "harm," cannot be meaningfully applied to it.

Such plausibility as the feedback model does have results from the fact that the processes of feedback mechanisms do have value-significance for the men who design, construct, and employ them. The functioning of the mechanism is successful or unsuccessful in terms of the goals and purposes of man. But, as we have seen, in judging the goal-directedness of behavior we are ultimately concerned only with the value-significance of the behavior's effects for the *agent*. Hitting an enemy ship is not the goal of the *torpedo*, although it is the goal of the men who launch the torpedo. The torpedo, not being alive, can have no goals.

It would, of course, be possible to accept the cybernetic meaning of "goal" and other teleological concepts, and simply stipulate that value-significance is not required for a process to be goal-directed. In this way, the altered concept of "goal" could be applied to the behavior of machines and inanimate objects in general. This, however, would only sidestep the problem of teleology, rather than solve it. The distinction between final states that are valuable to the agent and those that are not would still exist, and the importance of this distinction would soon force the problem to be reformulated in different terms.

The failure to identify the issue of value-significance has led many writers (e.g., Nagel) to regard the action of a furnace equipped with a thermostat as essentially similar to the action of the homeostatic temperature regulating mechanisms in the human body. It is often argued that our use of teleological explanation for homeostasis must justify its use in the case of the thermostat-equipped furnace, since, it is claimed, there is no relevant difference in the two types of action: each mechanism maintains a constant temperature level by the operation of built-in feedback mechanisms.

This very example, however, serves to pinpoint the crucial difference between a mere result and a goal. The alleged "goal" of the furnace's action has no value-significance to the furnace. The furnace is not alive; its existence is not conditional upon its action, it has no needs, and thus it makes no difference whether its "goal" is set at 70°, 50°, or 100°. But in the case of homeostatic temperature regulation in man, the goal is not an arbitrarily selected result: the temperature of 98.6° is necessary for man's survival, a mere 10% deviation spelling death. This is why it makes sense to say that man's homeostatic mechanisms operate *for the sake of* maintaining the temperature of 98.6°, while it makes no sense to say that a furnace operates for the sake of maintaining the temperature at which the thermostat is set (except, of course, in terms of the value-significance of that temperature for human beings).

The essential feature of homeostasis is not that it maintains a *constant* temperature, but that it maintains *the temperature level required for survival.* If human survival required a specific kind of temperature variation (e.g., higher when active, lower when resting), then the maintenance of a constant temperature level would represent a *failure* of the bodily mechanisms to fulfill their function. Obviously, success and failure in goal-directed action is judged relative to the survival need which the goal is to satisfy—not by whether or not some arbitrarily selected end state is reached.

The phrase "action *for the sake of* obtaining X" best emphasizes the implicit value-significance of the goal of a teleological process. We almost never use this phrase without understanding that the object of the action is something regarded as beneficial to the agent. "Action for the sake of obtaining X" means roughly: "action for the benefit that can be realized from obtaining X." For instance, "He did it for her sake" could be restated as: "He did it to benefit her." For this reason I consider this phrase, and its shorter version "action for X" to be paradigm cases of teleological language. It is interesting that the Aristotelian term *to hou heneka*, usually translated as "final cause," means literally: "the for what" or "that for the sake of which."[19] In teleological processes the goal is identified as a goal partly by virtue of the *benefit* which it confers on the agent.

Our sense that a similarity exists between vegetative and purposeful action has been partially explained by the fact that the ends of each have a value-significance to the agent. If conscious purposes were operative on the vegetative level, we would expect just the kind of action to be performed purposefully which is in fact performed automatically. If a plant had desires, it would be natural to assume that it desired to live and that it would consequently pursue those conditions which favor its survival. Vegetative actions proceed in just the way we would expect them to proceed if they were under conscious direction.

In contrast, non-teleological processes (or, those ordinarily regarded as such) do not have the kind of results we would expect from a consciously directed process. Even if, for example, a volcano were a conscious entity we can see no reason why it would want to erupt; the eruption does nothing *for* the volcano (in fact, since the volcano has no needs, the concepts expressing value-significance cannot be meaningfully applied here). Similarly, it might be possible to regard the actions of certain feedback mechanisms as *directed*, but since the element of value-significance is absent, we cannot regard their actions as *goal*-directed. Only when the result of an action has value-significance for the agent, can we see a real similarity to purposeful action.

VI

GOAL-CAUSATION

THE TELEOLOGICAL VS. THE ACCIDENTAL

We have analyzed living action from the standpoint of its cause (viz., self-generation) and of its effect (viz., value-significance). Still remaining is the essential issue: the relationship of the cause to the effect. For an action to qualify as teleological, it must not only reach an advantageous end state, the action must have been performed *because* it reaches that end state, That is, the goal must in some way serve as the *cause* of the action's occurrence and direction. Teleology implies *goal-causation.*

The fact that goal-causation is a necessary element in the analysis becomes clear when we consider its role in purposeful actions. The mere fact that a conscious action brings about a desired consequence does not in itself imply that producing that consequence constitutes the purpose for the sake of which the action was undertaken. Desired objects are sometimes obtained by accident.

For example, suppose a certain man holds a general desire to meet the kind of woman with whom he can fall in love. Imagine that while taking a bus to work he happens to meet such a woman, and this meeting leads to a happy romantic relationship. Taking the bus in fact led to the satisfaction of his desire, but we could not say he took the bus *in order to* satisfy that desire. His purpose in taking the bus was to get to his job—the meeting of the woman was just a fortunate accident.

An action's purpose is determined by reference to the desire or value that *caused* the agent to undertake the action.

Thus, a vegetative action that benefits an organism cannot be said to be directed toward the achievement of that benefit as a goal unless the benefit was in some sense a cause of the action, rather than occurring as a fortunate accident. Biologist George Williams makes this point in terms of the teleological concept "adaptation":

> A frequent practice is to recognize adaptation in any recognizable benefit arising from the activities of an organism. I believe that this is an insufficient basis for postulating adaptation and that it has led to some serious errors. A benefit can be the result of chance instead of design. . . . Consider a fox on its way to the hen house for the first time after a heavy snowfall. It will probably encounter considerable difficulty in forcing its way through the obstructing material. On subsequent trips, however, it may follow the same path and have a much easier time of it, because of the furrow it made the first time. This formation of a path through the snow may result in a considerable saving of time and food energy for the fox, and such savings may be crucial for survival. Should we therefore regard the paws of a fox as a mechanism for constructing paths through snow? Clearly we should not.[1]

Similarly, Francisco J. Ayala, discussing teleological explanation, observes:

> Teleological explanations imply that the end result is the explanatory reason for the *existence* of the object or process which serves or leads to it. A teleological account of the gills of fish implies that gills came into existence precisely because they serve for respiration.[2]

The central problem of teleology is that of explaining this apparent reversal of cause and effect involved in teleological causation. Since the future event of goal-attainment does not exist until the action's completion—and may never exist, since the action may fail to achieve its goal—how can the benefit derived from attaining the goal be the cause or explanatory reason of the action's occurrence or direction? The solution to this paradox in the case of purposeful action was seen to lie in the causal role played by the agent's present mental content—to quote Braithwaite again:

> Teleological explanations of intentional goal-directed activities are always understood as reducible to causal explanations with intentions as causes; to use the Aristotelian terms, the idea of the "final cause" functions as the "efficient cause."[3]

In the case of vegetative actions, however, there are no ideas, intentions, desires or other mental phenomena to which to appeal in resolving the paradox of "final causation." The crucial requirement of an adequate defense of teleology is to explain how the value-significance of a goal can cause the action directed toward that goal, without implying either that consciousness is involved or that possible future events can causally influence the present.

One task which the teleologist need *not* meet in defending goal-causation, is that of providing a general, philosophic explication of the concept of "cause." He need only compare the type of causation in purposeful and vegetative action. The problem of teleology is not one of defining the causal relation itself, but rather in identifying what kind of *objects* are involved in the causal relation in these cases. For instance, in discussing purposeful action, the question of how or in what sense a mental state can be a cause of a bodily action is entirely separate from the question of what renders purposeful action teleological. If, as Braithwaite suggests, the idea of the goal is an *efficient cause* in purposeful action, then the task of the teleologist is to discover what *efficient cause* in vegetative action explains its apparent future-causation, and he must do so without appealing to ideas or other mental phenomena.

Nor is it necessary that the form of causation in purposeful action and vegetative action be exactly the same—clearly they are not identical (the differences will be outlined subsequently). But to defend vegetative teleology, one must show that in vegetative action there is some genuine, objectively determinable, efficient causation by the goal. The value to the organism of attaining the goal must in some way promote or facilitate the action that attains it, and the requirements of attaining the goal must in some way control the onset and course of the action performed.

FUTURE ENDS AS BASED ON SIMILAR PAST ENDS

A. *In purposeful action*

In order to approach the problem of final causation in vegetative action, we should take a closer look at how this problem was solved for the case of purposeful action. In purposeful action, the attribution of causal agency was shifted from the future goal to some present mental content. But how are we to understand that present mental content? To what is it a response? In the case of normal sensory perception the mental content, the experience, is a response to the physical stimulation provided by some existing object. When we visually perceive an apple, for instance, the mental content is understood as being a response to the apple—a response caused by the light it reflects. In perception, the mental content is the product of the interaction of some environmental object with the subject's sense organs and nervous system.

But the mental content responsible for a man's performance of a purposeful action does not lend itself to this kind of analysis. A desire for a goal is not the product of an interaction of the goal with the agent's sense organs and nervous system. The desire for the goal is not a response to physical stimulation by the goal.

At least two considerations support this conclusion. First, in the case of purposeful actions which fail, the goal never comes into existence and consequently cannot be the source of the desire for the goal which causes the unsuccessful action to be undertaken. Second, we must distinguish in this context between the goal and the goal-object. For instance, when a hungry dog perceives a bone, the bone may serve as the goal-object for the ensuing behavior, but it is not the goal. The goal is to get and eat the bone. More specifically, the goal is generally some realization of the object's potential value-significance to the agent. The goal *qua* goal, then, is ordinarily not an object at all, but rather the activity of consuming or utilizing an object.[4] The significance of this point is that the consumption or utilization of the goal-object is necessarily a *future* event which accordingly cannot be the source of the present desire for the goal. Moreover, in many cases even the goal-object does not yet exist, or exists in a location or condition rendering it incapable of yet affecting the agent. A man who desires to travel from New York to San Francisco is not being physically stimulated by San Francisco. Both points show that in purposeful action one's desire cannot be a response to the goal. To what then is the desire a response?

The same question arises concerning the element of anticipation. If consciously held goals are described as "ends-in-view," it must be recognized that "view" is used metaphorically, since literally a *view* of a *future* end is impossible. What, then, is the actual sense in which we have a "view" of a future end?

On the basis of a general empiricist epistemology, answering this question is not overly difficult: any mental content which is not the effect of a direct sensory contact with existing objects is derivative from the material gained in past instances of direct perception. Specifically, our beliefs and desires about the future represent extrapolations from past instances of awareness of actual objects. A dog's desire for an affectionate pat from its master is a consequence of its memory of similar past expressions of affection.

As discussed in chapter 3, an animal's ability to act purposefully derives from its ability to learn to associate its performance of certain actions with pleasure or pain. In learned associations, the behavior results from a mental connection established between an object of a certain general type and the experience of pleasure or pain, such that any similar object re-evokes the former pleasure or pain. Young kittens, for example, are not afraid of fire, but once having been burned by a flame, they scrupulously avoid a second occurrence of that painful experience. The pain experienced in the original encounter with the flame has become permanently associated in the cat's mind with the memory of the flame itself, and anything recalling the memory of the flame also recalls the memory of the pain. In this manner, the cat *learns* to avoid anything that is perceptually similar to the original flame. It need not be assumed that the cat is aware of possible future harm that a flame could cause, but rather merely that the cat is remembering *past* hurt it actually experienced when it encountered an object perceptually similar to the present one. Even if animals cannot explicitly project the future, they can learn from the past.

Man, of course, is able to project the future *as* the future, rather than merely recalling the past, but this projection is still clearly based on past experiences (including the abstract, conceptual knowledge derived from those past experiences). A worker can desire to receive this week's paycheck because he received a paycheck last week (and, more to the point, because he has in the past been acquainted with instances of money, productive work, and exchange and has conceptualized these things).

Even a man who conceives of something new and unique does so by means of an imaginative re-arrangement of aspects of reality he has perceived in the past. When James Watt invented the steam engine he had to conceive of something which neither he nor others had ever encountered before, but that conception was possible only by virtue of his prior experience with steam and with machines. An artist can conceive of a painting which depicts objects unlike anything anyone has previously experienced, but he can do so only by means of a novel re-combination of the colors he has previously experienced through perception. This point is but an application of the empiricist dictum: "Nothing is in the mind that was not first in the senses." This does not mean that the mind is unable to create, but only that it is unable to create *ex nihilo*—the materials out of which the mind creates, and projects a future goal, are furnished in the last analysis by prior sensory contact with external objects.[5]

Thus, the future-oriented desires and beliefs characterizing purposeful action are to be understood as based upon the agent's past experience with relevantly similar previous goals and actions, given the agent's ability to associate or extrapolate.[6] This point has been made by Israel Scheffler in a discussion of teleological causation (although he applies this analysis more narrowly than I do). Scheffler offers the following explanation of the case of an infant that cries in order to attract its mother's attention:

> In the case of the infant, for example, it may be suggested that our "in order to" description of its present crying reflects our belief that this crying has been learned *as a result of the consequences of like behavior in the past*—more particularly, as a result of having in the past, received its mother's attention. Having initially cried as a result of internal conditions C, and having thereby succeeded in obtaining motherly solace, representing a type of rewarding effect E, the infant now cries in the absence of C, and as a result of several past learning sequences of C followed by E. The infant's crying has thus been divorced from its original conditions *through the operation of certain of its past effects*. These past effects, though following their respective crying intervals, nonetheless precede the *present* crying interval which they help to explain. The apparent future-reference of a teleological description of this present interval is thus not to be confused with prediction, nor even with mention of particular objects in the current environment toward which the behavior is directed [since the goal-object may be non-existent]. Rather, the teleological statement tells us something of the genesis of the present crying and, in particular, of the prominent role played by certain past consequences in this genesis. Such an account is perfectly compatible with usual conceptions of causal explanation, and, though sketchy and hypothetical, it indicates why goal-objects may very well be missing in some cases for which teleological explanation is nonetheless appropriate [my emphasis].[7]

In purposeful action, the future goal, being as yet non-existent and perhaps never to exist, is *not* a cause of the action, nor is the future goal the cause of the goal-idea in the agent's mind. The cause of the goal-idea, and hence of the action, lies in the psychological effects of *past instances of the goal*. (In the case of novel goals conceived by human beings, the cause of the goal-idea is to be found in the psychological effects of the previously perceived constituents of the novel goal.) Purposeful actions are possible because of the agent's ability to capitalize upon past experiences with similar previous actions and similar previous ends.

In purposeful action, the agent's desire for some anticipated result is the proximate cause. But since both that desire and the belief in the action's efficacy result in turn from experiences with similar previous objects reached by similar previous actions, a remote or ultimate cause is the value-significance of the *past* instances of the goal.[8]

B. In vegetative action

Can the same kind of reduction of future ends to past ends be accomplished for vegetative actions? To begin answering this question, we need to identify the proximate and ultimate causes of vegetative actions.

What are the proximate causes of vegetative actions? There are several. First, vegetative actions are self-generated—which, as we saw in chapter 4, implies the existence of both an internal store of energy, the "fuel," and a directive mechanism controlling the utilization of that energy. Second, the organism must possess the physical organ or structure for performing the action in question.

For example, in the case of the heartbeat, the fuel is the glycogen (and its derivatives) and the oxygen which jointly supply the chemical energy for the heart's contractions; the directive mechanisms include the pacemakers, the nerves which innervate the heart, and, ultimately, the entire nervous system; the organ for performing the action is, of course, the heart itself. We may simplify matters by treating the organ and the directive mechanism together as a unit, which I will refer to as "the mechanism" for the action. Thus, the mechanism for the heartbeat is the heart, as controlled by its pacemakers and nervous inputs.

The two main proximate causes of vegetative actions, then, are the fuel and the mechanism. If we were to apply this analysis to the operation of an automobile engine (ignoring, for the purposes of the example, the differences between the operation of such an engine and the vegetative actions of living organisms), we would say that the fuel is the gasoline plus the oxygen taken into the carburetor, and the mechanism is the physical structure of the engine (including its equivalent of directive mechanisms: the accelerator pedal and linkage, the distributor, the camshaft, etc.).

In addition to the fuel and the mechanism, another proximate cause is the triggering stimulus. As explained in chapter 4, a stimulus may be said to "trigger" a process when it activates the process but does not supply the energy for carrying out the process itself. The basic heartbeat, for example, is triggered by an electrical discharge originating in the sino-atrial node, with changes from that basic rate being triggered by input from the vagus nerve.

In the case of purposeful action, the triggering stimulus not only interacts physically with the organism, but also is perceived by the organism. The dog's action of approaching a bone is triggered by its perception of the bone. The child's crying is triggered by some stimulus (perhaps internal) which recalls the memory of the motherly solace it has learned to associate with crying. Even in conceptually guided human behavior of the most abstract sort, some stimulus is presumed to exist to trigger the chain of thought which motivates the action.

On the vegetative level, the stimulus is able to trigger the action because of the way the mechanism for the action is organized. The mechanism has certain *terms of operation* dictated by the nature of its directive mechanism(s). The way in which the mechanism is organized determines what will or will not trigger its behavior. Returning to the automobile analogy, because of the way the automobile engine is organized, pressure applied to the accelerator pedal will increase the engine's speed, but equal pressure applied to, say, the dashboard will have no such effect.

Since the triggering stimulus is generally provided by some external factor (e.g., sunlight in the case of phototropism), its causes will generally be located in some earlier external factors (e.g., nuclear activity inside the sun) that are irrelevant to teleological causation. Consequently, although the causes of the triggering stimulus do constitute indirect (or ultimate) causes of any given action, we will not be concerned with this particular causal chain. We are interested only in those ultimate causes which might constitute past instances of the action's value-significance (just as, on the purposeful level, we are not interested in what caused the dog's bone to be present).

To be an ultimate cause of an action is to be the cause of the proximate cause of that action. (It should be noted that in speaking of a cause as "ultimate" I do not mean that the cause is itself primary or irreducible—the cause is "ultimate" only relative to the proximate cause, not in an absolute sense.) In vegetative action the ultimate cause we are seeking is specifically the cause of the presence of the fuel and the mechanism which directly account for the action's occurrence.

At this point the following objection might be raised: in explaining the action by reference to the fuel, the mechanism, and the triggering stimulus, we have done enough; the explanation is complete—there is no need to raise the question of the causes of the proximate causes. Since we have fully explained the action by reference to its proximate causes, to ask "What caused the proximate causes?" is to shift to a new question, not to add anything to the answer to the original question of what caused the action.

There are several issues involved in this objection. The first concerns what constitutes a *complete* explanation? The explanation of a particular action is accomplished by identifying the causal factors that produced it. In a limited sense, any given vegetative action has been completely explained when its proximate causes have been identified—in fact, if it is not complete, this shows only that mention of some proximate cause has been omitted, perhaps due to the limitations of our current knowledge of the relevant physical laws. In this sense of "complete explanation," any incompleteness must be remedied by a closer investigation of proximate causes—*not* by appeal to ultimate causes or teleological laws. (Here I differ from vitalist teleologists, like Schubert-Soldern, who maintain that the explanation of vegetative action in terms of its proximate causes can never be complete in this sense, unless reference to a teleological "principle" or "entelechy" is included.)

To say the explanation of the action is complete in this sense, however, is not to say that going to the next level—explanation of the proximate causes themselves—can add nothing to our understanding of the action. In many cases the same event can be explained on different levels, and the appropriate level is judged relative to the kind of question to which the explanation is an attempted answer.

If, for example, the patient of a psychiatrist commits suicide by jumping off of a bridge, and the psychiatrist is asked: "Why did he jump?"—it would be ludicrous for him to answer: "Because his leg muscles received electrical stimulation from his brain." That answer might be satisfactory if the question were asked of a physiologist (*qua* physiologist), but in the present context the answer is absurdly narrow. The point of the question was clearly to determine what caused the patient to initiate that electrical stimulation of his leg muscles—specifically, to determine his motive for ending his life. The physiological answer, when given by the psychiatrist, may be said to be contextually incomplete. Contextual completeness is determined relative to a given level of inquiry.

The objection to seeking an explanation of the proximate causes perhaps arose as a reaction to the view that explanation of events requires an infinite series of explanations in terms of causes of causes of causes of. . . (This view is often assumed, for example, in arguments attempting to prove the existence of a First Cause.) But this (false) view of explanation is not entailed by my position. I am not maintaining that an explanation of a vegetative action in terms of its proximate cause is no explanation at all—nor that an explanation is not *valid* until the proximate cause is also explained. My position is that *both* explanation in terms of proximate causes and explanation in terms of ultimate causes are valid. There is a perfectly legitimate context in which the explanation of vegetative action in terms of its fuel, mechanism, and triggering stimulus is fully complete (e.g., for the science of physiology); yet this same explanation may be incomplete from the standpoint of another type of inquiry. As to the First-Cause approach to explanation, William Dray's response is accurate:

> A person who adopts the policy of always refusing to accept X as the explanation of Y unless the X itself is explained, begins to empty the term 'explanation' of its normal meaning. And if he goes on to demand that any explanation of a Y in terms of an X should at the same time explain X (and so on, *ad infinitum*) he empties the term of all meaning.[9]

If I can show that an ultimate cause of vegetative action is the value-significance of past instances of the goal, this will make vegetative action profoundly similar to purposeful action, even though an explanation of vegetative action in terms of its proximate (mechanical) causes is also fully valid.

The second part of my answer to the objection consists in the point that there is, in fact, a context in which explanation of vegetative action in terms of its proximate causes *is* insufficient. This is the context of evolutionary biology. The evolutionary biologist accepts the physiologist's and the biochemist's explanations of vegetative action in terms of its proximate causes, but is curious about the rather amazing fact that those proximate causes exist. The existence of such a highly organized mechanism as the heart *is* amazing when compared to what exists in the inanimate realm. Biochemist Albert Lehninger states:

> From the standpoint of thermodynamics the very existence of living things, with their marvelous diversity and complexity of structure and function, is improbable. The laws of thermodynamics say that energy must run "downhill," as in a flame, and that all systems of atoms and molecules must ultimately and inevitably assume the most random configurations with the least energy-content. Continuous "uphill" work is necessary to create and maintain the structure of the cell. It is the capacity to

extract energy from its surroundings and to use this
energy in an orderly and directed manner that distinguishes the living human organism from the few
dollars'. . . . worth of common chemical elements of
which it is composed.[10]

What is meant by saying that existence of living things
with their complex structures is "improbable"—a description
one encounters repeatedly in biological writings? Obviously
this description cannot be taken to mean that the existence of
complexly organized living systems is statistically rare, since it
is found in each of the cells of the billions of organisms cover-
ing the face of the earth. This organization is judged as
"improbable" relative to what would be expected in terms of
the laws of physics and chemistry *as they apply to inanimate matter.* In other words, the existence of the organization exhibited by cells, and *a fortiori* by complex organs composed of cells
such as the heart, is taken for granted by the sub-branches of
biology (e.g., physiology), but constitutes a problem to be
solved for the science of biology as a whole in relation to the
physical sciences.

The physiologist *qua* physiologist does not puzzle over the
high degree of organization characterizing the heart, the
liver, or the kidney, but the biologist, being concerned to
explain the basic nature of life—how it arose from inanimate
matter, how it diversified, proliferated, increased in complexity—does regard this organization as "improbable," puzzling,
and problematical. Simpson maintains:

> The point about explanation in biology that I should par-
> ticularly like to stress is this: to understand organisms
> one must explain their organization. It is elementary
> that one should know what is organized and how it is
> organized; but that does not explain the fact of the
> nature of the organization itself. Such explanation
> requires knowledge of how an organism came to be orga-
> nized and what functions the organization serves. Ulti-
> mate explanation in biology is therefore necessarily evo-
> lutionary.[11]

The following passage is from Dobzhansky's authoritative summary of current evolutionary biology, *Genetics of the Evolutionary Process*:

> In biology nothing makes sense except in the light of evolution. It is possible to describe living beings without asking questions about their origins. The descriptions acquire meaning and coherence, however, only when viewed in the perspective of evolutionary development.[12]

The question "what causes plants to turn toward the sun?"—or more generally: "What causes this vegetative action to occur?"—can be answered on different levels, corresponding to explanation in terms of "final cause" and explanation in terms of efficient cause. But explanation in terms of "final cause" is not to be understood as explanation in terms of a different sense of "cause," but as explanation in terms of a different kind of object participating in the causal relationship. The traditional notion of "final causation," growing out of Aristotle's doctrine, views efficient cause and final cause as if they were two separate forces acting on the organism: the efficient cause pushes the organism from behind and the final cause pulls it from in front (which would imply the actual existence of future goals serving as the source of this pull). The view I am defending, on the other hand, *assigns causal efficacy only to efficient causes*, but distinguishes between two kinds of efficient cause: proximate and ultimate.

According to this view, a causal chain is teleological when the value-significance of a certain kind of object functions as an ultimate efficient cause of an organism's action—specifically, the kind of object required is one that has the same kind of value-significance to the organism as does the future object that is regarded as that action's goal.

This analysis is readily applicable to those conscious actions we regard as purposeful. Take, for instance, Scheffler's example of the child who cries with the purpose of gaining motherly solace. The action is purposeful because its efficient cause is the child's desire to obtain the solace. This desire, in turn, stems from the child's learned association of performing the action of crying with obtaining motherly solace. The action is therefore teleological by the above analysis: the value-significance of an item of a certain kind (the pleasure derived from past instances of motherly solace) is an indirect efficient cause of the child's behavior—and, specifically, the item has the same kind of value-significance as does the future object that is regarded as its goal.

Lest this analysis seem too abstruse to account for our ordinary sense that the child's behavior is purposeful, let's consider a very typical exchange between a father and mother:

> Mother: Junior is crying again.
> Father: Of course he's crying—you've spoiled him. Every time he's cried, you've rushed in to comfort him and make a fuss over him. He's crying because he's learned that that will get him your affection. It's all those times you've rushed in to comfort him that have made him a cry-baby.

What is the father saying, if not that it is the past instances of obtaining the solace by crying that have caused the child to associate crying with solace, and which thus explain his present attempt to achieve that goal? The father is not attempting to explain the child's action by assigning causal agency to the *future* goal—in fact, the purpose of his remark is to prevent the occurrence of that future goal by persuading the mother to ignore the child. Rather, the father is explaining the present behavior by reference to the indirect causal role (in the sense of efficient causation) played by past instances of the goal.

Accordingly, to maintain that vegetative actions are teleological is to hold that vegetative actions are likewise ultimately caused by the value-significance derived by the organism from prior instances of the object that is regarded as its goal. This formulation again reduces "final causation" to indirect efficient causation—efficient causation by past instances of the goal.

In summary, the direct efficient causes of vegetative

actions are the fuel, the mechanism, and the triggering stimulus. The relevant indirect or ultimate efficient cause of these actions would, then, be the efficient cause of the fuel and of the mechanism. If vegetative action is to be legitimately classed as goal-directed, the kind of indirect efficient cause required is one having the same kind of value-significance to the organism as does the future object regarded as the action's goal. Since "value-significance" on the vegetative level means "survival-significance," this becomes the requirement that the indirect cause of a given vegetative action have the same kind of survival-significance as the future goal.

Putting all these points together, we can say that a vegetative action will qualify as teleological if it can be shown to be a self-generated action caused by a mechanism whose existence, organization, fuel, and terms of operation result from the survival benefit that past instances of the goal have provided the organism in similar previous circumstances. In such a case, the action would qualify as teleological in that its ultimate (efficient) cause would be the value-significance of (past instances of) its goal.

The only remaining question is: do vegetative actions in fact satisfy those conditions? An affirmative answer to this question is entailed by a special consequence of the conditional nature of life: the principle of natural selection.

VII

GOAL-CAUSATION
AND NATURAL SELECTION

Since Darwin, biologists have recognized that virtually all of the characteristic structures and functions of living organisms are adaptations—i.e., items which evolved because of some contribution they make to the organism's survival. Adaptations are (heritable) features whose survival value has caused organisms possessing them to be naturally selected in evolution. This, I will argue, amounts to a *teleological* explanation of those structures and functions.

The proximate cause of any vegetative action lies in the physical make-up of the organism. The proximate cause is the organism's possession of the fuel and the mechanism which physically determine it to act in a given manner when triggered by a given stimulus. But the explanation of the existence of an organism possessing both that fuel and that specific, "improbable" mechanism lies in the survival value of the action they determine. The ultimate efficient cause of the existence of organisms possessing adaptive mechanisms is the survival value gained by the actions which these mechanisms have made possible. A mechanism is "adapted" to survival when its existence is due to the survival value of the action it determines. (The "survival value" of a trait is its capacity to facilitate the perpetuation of organisms possessing it, in a given environment.)

Francisco Ayala states:

> In *The Origin of Species* Darwin accumulated an impressive number of observations supporting the evolutionary origin of living organisms. Moreover, and perhaps most importantly, he provided a causal explanation of the evolutionary process—the theory of natural selection. The principle of natural selection, as Darwin saw it, makes it possible to give a natural explanation of the adaptation

of organisms to their environment. With *The Origin of Species* the study of adaptation, the problem of design in nature came fully into the domain of natural science.

Darwin recognized, and accepted without reservation, that organisms are adapted to their environments, and that their parts are adapted to the functions they serve. Fish are adapted to live in water, the hand of man is made for grasping, and the eye is made to see. Darwin accepted the facts of adaptation and then provided a natural explanation of the facts. One of his greatest accomplishments was to bring the teleological aspects of nature into the realm of science. He substituted a scientific teleology for a theological one.[1]

The substitution of a scientific teleology for the earlier supernatural or animistic conceptions was, however, only implicit in Darwin's work. The teleological significance of Darwin's principle of natural selection has only recently begun to be made explicit. Generally, interest in natural selection has been focused on its role in explaining the adaptiveness of the *structural features* of organisms. The *actions* of organisms, however, are equally as adaptive as their structural features. (In fact, since each structural feature represents the product of the ontogenetic process of development, any such feature can be viewed as the goal of that self-generated process of development.) And just as natural selection explains the adaptiveness of structure, it explains the adaptiveness of actions: those actions which have reached end states possessing survival value tend to be repeated; those actions that fail to reach end states possessing survival value tend to be eliminated by virtue of the higher mortality rate of organisms performing them (including the relative inability of such organisms to replace themselves through the production of viable offspring).

Natural selection is usually looked upon as a principle explaining evolutionary change, but it may also be looked upon as explaining the maintenance or perpetuation of features as well. A basic premise of natural selection is what I have called the "conditional" nature of life—i.e., the fact that the continued existence of any living organism requires its successful performance of specific kinds of self-generated action in attaining specific ends. Not to act is to die; to act unsuccessfully is to die. This is at least part of what Darwin had in mind by the phrase "the struggle for existence."[2] The action required is not *primarily* the avoidance of external threats, but rather the positive achievement of the items that the organism can utilize for self-maintenance. Living organisms, George Wald observes, "need a constant supply of material and energy to maintain themselves. . . . When, for want of fuel or through some internal failure in its mechanism, an organism stops actively synthesizing itself in opposition to the processes which continuously decompose it, it dies and rapidly disintegrates."[3]

Not only is survival conditional upon action, but also the reverse is true: action is conditional upon survival. The sense in which survival is conditional upon action is that explained above: failure to act successfully leads to death. The sense in which action is conditional upon survival is equally straightforward: dead organisms cannot act.

This reciprocal relationship between action and survival is the basis of the kind of selection which accounts for the existence of teleological causation on the vegetative level: unsuccessful actions tend to be eliminated by the death of the organisms "programmed" to perform them—the dead organisms can neither repeat the unsuccessful action itself nor pass on the genetic "programming" for that action to other organisms (since dead organisms cannot perform the action of reproduction). Conversely, successful action qualifies as successful by virtue of its ability to provide the organism with the means to repeat that action in the future and to pass on the genetic "programming" for that action to its offspring, which will thus repeat the action in the new generation as well. In this sense, to explain an action as the product of natural selection is to explain it in terms of the causal role played by the results of previous instances of that action. *An action that has been naturally selected is one whose re-occurrence is due to its past success in contributing to the survival of the organism.*

All vegetative actions, with very few exceptions, occur only because of the survival benefit those actions provide the organism in the kind of environment to which it is adapted. In other words, the vegetative actions of organisms are normally goal-directed, with survival as their ultimate goal.

Vegetative actions are explained teleologically by reference to the ultimate (efficient) causal role of the survival benefit the organism has derived from the end states of past instances of that action. It is the very survival contribution these actions make that causally explains why such actions continue to be performed: a necessary condition of the reoccurrence of these actions is the survival of the organism; in turn, a necessary condition for the survival of the organism is the survival contribution made by past instances of these actions. Thus, an indirect necessary condition for the reoccurrence of these actions is their past contribution to the survival of the organism. Here, then, is the goal-causation we have been seeking: these actions are goal-directed in the specific sense of being ultimately caused by the value-significance of past instances of the object that is regarded as their goal.

To adapt Braithwaite's formulation: in vegetative action a *past instance* of the "final cause" functions as the "efficient cause."

The significance of natural selection for the issue of teleology has also been identified by Ayala. In an article published in 1970 under the title "Teleological Explanations in Evolutionary Biology," Ayala argues:

> Teleological mechanisms in living organisms are biological adaptations. They have arisen as a result of the process of natural selection. The adaptations of organisms—whether organs, homeostatic mechanisms, or patterns of behavior—are explained teleologically in that their existence is accounted for in terms of their contribution to the reproductive fitness of the population.[4]

Ayala's theory differs from the one presented here in at least one respect: he holds that the ultimate goal of teleological behavior is not the survival of the individual organisms, but the production of viable offspring—"reproductive fitness," in the terminology of modern genetics. "The ultimate end to which all other functions and ends contribute is increased reproductive efficiency."[5] The general relation between individual survival, reproductive success, and group survival will be discussed in detail in chapter 9. Secondly, it is not entirely clear whether Ayala regards teleological explanation to be a form of *causal* explanation; he seems, rather, to regard teleological explanation as answering an irreducibly distinct kind of question—"what for?" rather than "by what agency?"[6]

In addition, several authors have adumbrated the thesis presented here. In 1967 Dobzhansky wrote:

> Body structures and functions that are formed fit together not because they are contrived by some inherent directiveness named "telos," but because the development of an individual is part of the cyclic (or, more precisely, spiral) sequence of development of the ancestors. Individual development seems to be attracted by its end rather than impelled by its beginning; organs in a developing individual are formed for future uses because in evolution they were formed for contemporaneous utility.[7]

Pittendrigh (1958) states that the question underlying teleology (or "teleonomy" as he prefers to call it) is:

> "What is the origin of that information which underlies and causes the organization [of any given mechanism]?" The second law of thermodynamics exacts, as it were, a general payment from the universe in the form of a loss of information, an increase in disorganization; and we might accordingly restate the question as follows: "How has the information content of the genotype accumulated in the face of the universal tendency to maximize entropy?" The only general answer to this question is that outlined by Darwin: natural selection. Selection, as [R. A.] Fisher put it, is a device for generating a high degree of improbability. And, accordingly, it is by understanding the nature of natural selection that we get our best insights into the nature of the adaptive organization

it generates.[8]

Finally, Simpson has also defended a similar view (1947):

"What for?"—the dreadful teleological question—not only is legitimate but also must be asked about every vital phenomenon. In organisms, but not (in the same sense) in any non-living matter, adaptation *does* occur. Heredity and muscle contraction do serve functions that are *useful* to organisms. . . . It is still scientifically meaningful to say that, for instance, a lion has its thoroughgoing adaptations to predation *because* they maintain the life of the lion, the continuity of its species, and the economy of its communities.

Such statements exclude the grosser, man-centered forms of teleology, but they still do not necessarily exclude a more impersonal philosophical teleology. A further question is necessary: "How does the lion happen to have these adaptive characteristics?" or, more generally and more colloquially, "How come?" This is another question that is usually inappropriate and does not necessarily elicit scientific answers as regards strictly physical phenomena. In biology it is both appropriate and necessary, and Darwin showed that it can here elicit truly scientific answers, which embody those that go before. The fact that the lion's characteristics are adaptive for lions has caused them to be favored by natural selection, and this in turn has caused them to be embodied in the DNA code of lion heredity.

Several pages later, Simpson draws the following conclusion:

The teleonomic aspect is involved in the explanation of genetic information in organisms by adaptive processes, notably those of differential reproduction, or natural selection in the modern sense.[9]

None of these authors, to the best of my knowledge, have explicitly made the point that natural selection occurs by virtue of the survival value of *past instances of the goal*; and none have, therefore, gone on to defend a concept of "goal-directed action" defined in terms of the directiveness supplied by past instances of the goal. (Nor am I implying that they would agree with those further points, as I am presenting them here.) Perhaps the clearest understanding of the relationship between past ends and future ends is found in the statement of George Williams (1966):

> Only in an endlessly recurring cycle, as is shown by the succession of generations in a population, can one class of events be both the cause and the effect of another.[10]

Let us now examine the three typical examples of vegetative action to see just how in each case the kind of goal reached explains the occurrence of the action by explaining the organism's possession of the fuel and the mechanism for that action.

CELLULAR RESPIRATION

The goal of cellular respiration is the production of utilizable energy, in the form of the ATP molecule. The survival value of ATP is direct and all-encompassing. A recent textbook of biology states:

> ATP is often called the universal energy currency of living things, and the characterization is fully justified. It is energy stored in the energy-rich phosphate bonds that is used to do all manner of work: synthesis of more complex compounds, muscular contraction, nerve conduction, active transport across cell membranes, light production, etc. Whenever any organism is doing work of any kind, you can be certain that ATP is involved; the energy price of the work is paid in ATP's (i.e., the energy is obtained through the breaking down of ATP to ADP.)[11]

ATP, as the goal of respiration, has survival value by virtue of the fact that it fuels all the processes that keep the organism alive. Without the ATP provided by respiration, the organism would disintegrate. But respiration itself is one of the processes which organisms perform. This means that respiration fuels itself: one consequence of successful respiration in a healthy cell is continued respiration. Specifically, respiration is the oxidative breakdown of glucose; glucose is the fuel that is "burned" in respiration. By what means does the organism obtain this fuel? Glucose is obtained either by the process of photosynthesis (in the case of autotrophic organisms) or by feeding on glucose-containing organisms (in the case of heterotrophic organisms). In each case the energy required to obtain the glucose is provided by ATP (ATP is required for photosynthesis and for all the actions involved in food acquisition and digestion by heterotrophs).

Thus, present acts of respiration are made possible by the ATP produced by past acts of respiration. From the standpoint of its required fuel, respiration is a process which reoccurs only because of the survival value that its goal—ATP—provides.

We can look at this same process from the standpoint of its mechanism as well as its fuel. Biochemically, respiration is initiated and directed by a specific type of sub-cellular mechanism: the mitochondrion. Viewed as a mechanical process, respiration is caused by the immediate presence of certain chemical compounds in a certain spatial arrangement (determined by the spatial organization of the mitochondrion itself) in the presence of certain catalyzing enzymes. The process is not mystical—the mitochondria act as physical-chemical mechanisms according to the standard laws of chemical affinity and thermodynamics.

The structure of the mitochondrion is part of the proximate cause of respiration—but what is the cause of this proximate cause? How is it that the cell possesses functioning mitochondria? Mitochondria are, like all the complex subcellular organelles, very fragile and "improbable" systems. How is it that the mitochondria do not disintegrate, as they do in dead cells? In answering this question, the mechanistically minded biochemist will simply make reference to the various mechanical processes of self-maintenance and self-repair which preserve the integrity of the mitochondria. But these processes are themselves driven by ATP—the ATP produced by past acts of respiration. In giving a causal account of the continued existence of the mitochondria we must make reference to the survival contribution made by the ATP produced in past instances of the functioning of the mitochondria. The mechanism for respiration continues to exist only because it is "a means to an end"—i.e., only because it leads to the production of ATP.

It is in this sense, then, that cellular respiration may be termed a "goal-directed action": past instances of its goal (ATP) are, by virtue of their survival value, indirect causal agents in the re-occurrence of that action. ATP is an indirect cause of respiration both in supplying the energy for the procurement of the direct fuel (glucose) and in supplying the energy for the maintenance of the direct mechanism (the mitochondrion).[12]

PHOTOTROPISM

The goal of phototropism is increased absorption of sunlight. This goal has survival value in that sunlight drives the process of photosynthesis.

As discussed in chapter 4, phototropism is essentially an asymmetrical growth rate in response to differential illumination. Thus, the same glucose that fuels the plant's growth is *ipso facto* the fuel for phototropism. This glucose itself is derived from the process of photosynthesis. The energy for photosynthesis is derived from sunlight (and from ATP which in turn owes its own energy content ultimately to sunlight). Consequently, phototropism is fueled indirectly by past instances of its goal: increased absorption of sunlight. Increased absorption of sunlight causes increased glucose synthesis, which through respiration provides extra energy to fuel the action of turning toward the sun.[13]

One might question whether the fuel for phototropism comes only from that *extra* sunlight which earlier phototropisms have provided. The plant would receive *some* sunlight even if it did not turn toward the sun. Perhaps phototropism is actually fueled by the plant's general store of glucose. If so, the existence of the fuel for phototropism would not have as a necessary condition the survival benefit derived from past instances of its goal (i.e., the extra glucose derived from the extra sunlight). It is possible, however, to reject this hypothesis: the evolutionary "justification" of phototropism is that the amount of glucose it produces *exceeds* the amount of glucose consumed as fuel in its performance. In order to have been selected in evolution, phototropism must cover its own energy costs and in fact be "profit-making" on net balance. If phototropism represented a drain on the plant's store of glucose—if it did not provide more glucose than it consumed—it would have been eliminated rather than perpetuated in natural selection. (This assumes, of course, that the adaptive value of phototropism lies exclusively in its addition to glucose production through increased sunlight absorption.) Accordingly, we can infer that the fuel for phototropism comes only from the *extra* glucose which its own past occurrences have provided.

Let us now discuss the explanation of the *mechanism* for phototropism. The phototropic mechanism consists of two elements: the general mechanisms for growth and the auxin mechanism for regulating the growth rate in accordance with the direction of incident sunlight. The following remarks are directly relevant to the auxin mechanism, but apply with equal force to the general mechanisms for growth insofar as they are specifically relevant to phototropism.

The explanation of the auxin mechanism in terms of the survival value of the phototropic action it determines is more complex than the preceding explanation of the mitochondrial mechanism in terms of the survival value of the action it determines. This is due to the fact that the extra glucose made possible by phototropism is probably not an absolute requirement of a given plant's survival in its lifetime; probably the survival significance of phototropism becomes decisive only after many generations. Even if phototropism provides but a slight advantage to plants performing it, this advantage will be translated into an expanded population of phototropic plants in later generations, and there will come a time when there will be offspring which could not have been produced but for the extra glucose from phototropism. Although no cell could survive without the ATP made possible by its mitochondria, we may assume that a single phototropic plant might well survive without the extra glucose made possible by phototropism. But, the advantage gained by that extra glucose will enable it to reproduce more successfully than varieties which lack the phototropic mechanism.

Furthermore, a survival *advantage*, a feature which merely facilitates survival, can become an outright survival *requirement* in the context of biological competition. In his introduction to *The Origin of Species*, Darwin outlines the role played by this competition:

> In the next chapter the Struggle for Existence amongst all organic beings throughout the world, which inevitably follows from the high geometrical rate of their increase, will be considered. This is the doctrine of Malthus, applied to the whole animal and vegetable kingdoms. As many more individuals of each species are born that can possibly survive; and as, consequently, there is a frequently recurring struggle for existence, it follows that any being, if it vary *however slightly* in any manner profitable to itself, under the complex and sometimes varying conditions of life, will have a better chance of surviving, and thus be *naturally selected* [emphasis added in the first instance].[14]

The resources required for any given type of organism exist in a limited quantity in its environment. For example, the growth of a plant requires fertile soil, but in any given environmental area there is only so much fertile soil to be had. Similarly, on any given area of land there is only so much incident solar radiation per square foot, and thus only a certain number of photosynthesizing plants can be supported. For example, in dense forests some types of shrubbery, which would otherwise be able to survive, are excluded because the leaves of the trees have out-competed the shrubs in the acquisition of sunlight.

In the intense competition of the plant world, any mechanism which enables a plant to appropriate more of the vital resources it needs can spell the absolute difference between life and death. A plant whose leaves are oriented perpendicularly to the sun's rays can steal sunlight from plants whose leaves are oblique to those rays. Any mechanism which can keep the plant's leaves oriented perpendicularly to the sunlight will accordingly afford that plant a better chance of staying alive. The extra sunlight provided by phototropism might not be necessary to the plant's survival in isolation (i.e., in an environment lacking competitors), but it becomes absolutely essential in the context of intense competition for limited resources. Under conditions of competition, a slight increase in fitness means a higher probability of survival as opposed to extinction. (The pioneering work of R. A. Fisher[15] has shown that even a very slight increase in the probability of survival, 1% or less, can cause a trait originally existing in just a single individual to spread throughout the entire population in only a few hundred generations.)

Sometimes the competition involved is not so direct as that pictured above; the competition often takes place through reproduction. A plant enabled by phototropism to produce extra glucose will have extra energy to expend in reproduction. This could mean the phototropic plant produces more seeds of the same size, the same number of seeds of a larger size, hardier seeds, or some combination of these conditions. In any event, its descendants will make up a greater percentage of the new population in the next generation (especially if the new types of plants can be established only by displacing older types within a given area). Likewise, the phototropic plants would make up an even greater percentage of plants in the third generation, and so on until ultimately only the phototropic plants would exist in that area.

It is important to realize that this reproductive competition still reduces to the direct form of competition for scarce resources discussed above. In the direct competition, the action of phototropism enables plants performing it to take sunlight away from non-phototropic competitors; in the reproductive competition, what is taken away might not be sunlight but ground space for germination. The extra sunlight provided by the phototropic mechanism indirectly causes the plants with the genes for that mechanism to produce more, bigger and/or hardier seeds, and thus these seeds physically displace some of the seeds of the non-phototropic plants.

The net result in this case is that a slight survival advantage provided by the phototropic mechanism results in the safe establishment of a greater number of phototropic offspring than would have been possible without that survival advantage. The additional phototropic offspring *owe their very existence* to the extra sunlight that the phototropic mechanism provided their ancestors in the previous generation. Thus the maintenance and proliferation of auxin mechanisms over time in the population is caused by the survival value of the action that mechanism determines.

Whatever competitive advantage in fact accounts for the evolution of the auxin mechanism, in explaining the presence of this mechanism in contemporary plants we must refer to the survival value it has for those plants which possess it. Without this evolutionary explanation—which amounts to a teleological explanation—the existence of the auxin mechanism remains unintelligible, an enigma, or "evidence" of the work of a cosmic Designer.

The mechanist may object that he can explain the existence of the auxin mechanism *without* making reference to the causal role of the survival contribution made by phototropism. He may explain the presence of the auxin mechanism as the product of reproduction. When asked "Why does this plant have the auxin mechanism?" he can simply response: "Because its ancestors did, given that reproduction is in kind." This answer, however, takes for granted that the reproduction was successful—a natural position when viewing the process after the fact. But, as we have seen, the reproductive success of the phototropic plants was greater than that of the non-phototropic variety, and may have, in fact, been achieved at the price of their reproductive failure. The question then arises: "Why have plants with the auxin mechanism been able to reproduce with greater success than similar plants without this mechanism?" The answer lies in the survival advantage provided by the acts of phototropism effected by the auxin mechanism.

Of course the mechanist may phrase his answer to this last question in terms that do not *explicitly* name that survival advantage as a survival advantage. He may answer by saying that the phototropic plants have a greater probability of successful reproduction because their seeds are hardier. When asked to explain why their seeds are hardier, he may simply refer to the chemical processes which underlie seed production, and when asked to explain the differences in these chemical processes between the phototropic and non-phototropic varieties, he may answer in terms of the differences in earlier chemical processes, and so on. Each link in the causal chain leading to the dominance of the phototropic plants will be explained in terms of its proximate efficient causes, without ever using terms like "survival value," "earlier instances of the goal," etc. Nevertheless, some of the proximate efficient causes, some of the physical-chemical processes, to which he *must* refer will be just those items which stand in the relationships designated by "survival value," "earlier instances of the goal," etc.

For instance, in explaining the increased hardiness of the phototropic plant's seeds, he will eventually have to mention the increased glucose possessed by the parent plant. In explaining that extra glucose, he will have to mention the extra sunlight obtained by the parent plant. In explaining the extra sunlight, he will have to mention the more nearly perpendicular orientation of its leaves to the incident sunlight. In explaining the orientation, he will have to mention phototropism and the auxin mechanism. Thus *whether he names them as such or not*, his explanation of the present existence of the auxin mechanism will have to include reference to the survival value of past instances of the goal as attained by past instances of the action determined by the auxin mechanism.

Whether he names it in these terms or not, the extra sunlight absorbed by the plant's ancestors *is* an earlier instance of the goal of the current plant's phototropism, the increased hardiness of the seeds *is* the survival benefit derived from that goal, and the increased hardiness *does* indirectly cause the present existence of plants possessing the auxin mechanism.

Thus the mechanical causal chain does add up to a teleological form of causation, although it is possible to refrain from adding it up and instead list each proximate mechanical cause in sequence. The point is that any mechanical explanation of phototropism sufficiently carried out will be teleological by implication, even if that implication is not made explicit.

The same is true of purposeful action: it is possible to give an explanation of purposeful behavior without naming "purposes," "desires," or "past instances of the goal" as such, even though the referents of these terms must be mentioned under some other labels in the explanation. For instance, in the case of a child who cries in order to gain motherly solace, the following explanation could be given: the child cries because it is responding to certain stimuli (including internal stimuli); those stimuli lead to the crying response because of the associative connection the child has formed between crying when those stimuli are present and obtaining relief from discomfort; the presence of that associative connection is explained by the manner in which the child's mind processed certain other stimuli in the past.

This type of explanation, which even includes references to mental states, and thus is not in a strict sense "mechanical," avoids describing the causal items in teleological terms, just as the mechanical explanation of phototropism does. Nevertheless, a teleological perspective on the same facts is possible here as well. In the explanation of the child's crying, the "certain other stimuli in the past" is also identifiable under the description "earlier perceived instances of the goal," the "relief from discomfort" is also identifiable under the description "the value-significance of the goal to the agent," etc. I see no relevant difference between purposeful and vegetative action with regard to the simultaneous possibility of teleological and non-teleological forms of explanation. The two forms of explanation are complementary—the difference being only one of level of abstraction.

THE HEARTBEAT

In a discussion of teleological causation, Arthur Pap states: "It would be odd to say that the bloodstream *causes* the beating of the heart. . ."[16] "Odd" or not, this is precisely the case. Specifically, past instances of the heartbeat's goal—i.e., past instances of blood circulation—cause the re-occurrence of the heartbeat in two ways.

First, with regard to fuel: the heart is a muscle, and as such its operation requires oxygen and glycogen in the same manner as does any other muscular contraction. These materials are supplied to the heart by the blood which flows through the coronary artery. The blood is pumped through the coronary artery by the heart itself. Thus the heartbeat causes the circulation of the blood which carries the fuel for the re-occurrence of the heartbeat.

In addition, from a wider perspective, we can ask how the fuel for the heartbeat got into the animal's bloodstream in the first place. In answering this question, we would trace back a series of actions such as digestion, alimentation, ingestion, and food-acquisition—all of which are possible only to an animal that is alive—i.e., all of which are made possible by the nutrients pumped by the heart through the circulatory system throughout the entire body of the animal.

If the animal's heart had not functioned in the past, its other muscles could not have functioned to obtain the food, its mouth could not have masticated the food, its stomach and intestines could not have digested the food, the glycogen could not have passed into the bloodstream, and finally the heart could not have continued to beat. Hence from both perspectives we see that the circulation of the blood in the past is a cause of the present heartbeat: past instances of blood circulation supply the fuel necessary for the heart's present contractions.

Second, with regard to its mechanism, the very survival of

the living cells composing the heart is sustained by the blood circulation which the heartbeat causes. According to a cardiologist, "the muscles of the heart cannot survive total deprivation of blood flow for longer than 30 minutes."[17]

Thus, the blood circulation is responsible not only for fueling the heart's action, but also for sustaining the "improbable" mechanism for producing the heartbeat—a mechanism which is irreparably damaged when deprived of blood circulation for only a half an hour: the heart itself. The thousands of deaths each year from coronary thrombosis are a tragic demonstration of the heartbeat's absolute causal dependence upon blood circulation.

There is no escaping the fact that the major vegetative actions of cells, plants, and animals are flatly necessary to survival, and hence can recur only if successfully performed. Animals with non-functional hearts, livers, kidneys, etc., cannot survive or reproduce. Hence the phylogenetic perpetuation of organisms with properly functioning organs is made possible by the successful functioning of those very organs. But even "minor" vegetative actions have survival significance, because, being self-generated, it takes energy to perform them.

> Every organism has available to it, at any stage of its life cycle, only a certain amount of energy with which it must maintain its life processes. . . . In addition to the supply of stored energy, the amount of "new" energy that the organism can obtain over a certain period of time is also limited. . . . The amount of energy may be very minute for some [life processes] such as breathing or sitting quietly and scanning the surroundings with its eyes, but every little bit adds up.[18]

Any energy investment that turns out to be unprofitable puts that organism at a disadvantage in relation to competitors that are programmed to invest their energy more profitably. Hence natural selection acts to fine-tune vegetative actions toward the most efficient use of energy, even when less efficient uses are not lethal, but simply hinder mean long-run survival.

Each of these three instances of vegetative action is thus amenable to teleological explanation in terms of the ultimate causal role played by the survival value of past instances of the goal involved. Although the proximate causes (the fuel and the mechanism) are mechanical, in each case the cause of the proximate causes—and thus the indirect or ultimate cause of the action itself—lies in the contribution that past instances of the goal have made to the organism's survival.

In contrast, consider a clearly non-teleological aspect of a vegetative action: the heart's production of a thumping sound. There is no survival value to this consequence of the heartbeat; the thumping sound makes no contribution to the survival of the animal and is not an indirect cause of its own re-occurrence. Consequently, by the present analysis we cannot say the production of the thumping sound is a goal of the heartbeat.

In the case of teleological causation, the relationship of action to goal may be summarized as follows: the action has value-significance in that it fulfills some survival need of the organism; hence, its ability to fulfill this need is an efficient cause of the continued survival of the organism; the continued survival of the organism makes possible its re-performance of that action at a later time. Or, looking at the same relationship retrospectively; the organism could not now perform the action if it had died; the organism would have died (or would never have come into existence) but for the physical contribution to survival made by past instances of the goal as attained by past instances of that action. An action may be regarded as teleological whenever it would not occur but for the survival contribution of past instances of its goal. As Ayala states in the passage quoted earlier:

> Teleological explanations imply that the end result is the explanatory reason for the *existence* of the . . . process which serves or leads to it.[19]

Whenever the *existence* of an action can be causally explained by reference to the nature of the kind of end result it has had in the past (via the ability of that kind of end result to account for the present existence of an organism equipped to perform that action) the action is teleological.

THE NATURE OF NATURAL SELECTION

Teleological causation is thus a consequence of natural selection. Natural selection is not merely the selection *of* processes having survival value, but also selection *by* that survival value: the perpetuation of beneficial measures is caused by the very benefits they attain. (In fact, as argued in chapter 5, this is the basis of identifying some result of a process as "beneficial.")

In this sense natural selection differs from what Darwin called "artificial selection." It is possible for man to select biological processes, rather than letting the processes select themselves, as it were. It is possible for man to choose to select processes which reach a given kind of end state independently of whether that end state is (otherwise) beneficial or harmful to the organism that acts.

For instance, suppose a geneticist has a population of fruit flies, some of which flutter their wings more often than others, due to differences in their genetic makeup. He might divide the population into "flutterers" and "non-flutterers" and arbitrarily decide to feed only the flutterers. In this case he has *artificially* selected frequent wing-fluttering; frequent wing-fluttering is selectively perpetuated not because of any intrinsic survival value it brings (we may even assume that in the natural environment this action represents a waste of the fly's energy resources), but solely because of the arbitrary choice of the geneticist to feed them.

(Of course, in a wider sense, the arbitrary choices of the geneticist still form part of the environment of those particular flies. Accordingly, artificial selection is not merely analogous to natural selection, it is actually a special case of natural selection. We single out "artificial selection" by reference to what has survival value only because of volitional human intervention.)

GENETIC SELECTION AND ONTOGENETIC SELECTION

The type of "natural selection" which grounds vegetative teleology is not *only* that form of selection to which biologists appeal in explaining the course of evolution. As Dobzhansky has noted, "natural selection is a common name for several cognate but distinct processes. . ."[20] The kind of "natural selection" that is relevant to teleology is: differential perpetuation of alternative forms of action. Whenever an action is perpetuated because of the survival contribution it has made in the past, that action has, in the sense explained, been *selected* for re-occurrence by its survival value: it owes its present existence (partly) to the kind of end state it brings about.

Consequently, in the teleological sense of "selection," selection of a vegetative action can occur within the lifespan of a single individual organism (ontogenetically). For instance, the present beating of my heart owes its existence to the survival contribution made by the results of my heartbeat during the last thirty minutes. In contrast, the kind of natural selection that evolutionary biologists ordinarily have in mind would *not* embrace the perpetuation of actions occurring within the lifespan of an individual organism. Evolutionary biologists are concerned mainly with the kind of selection explaining lasting evolutionary *change*, as in speciation. A mechanism for an action that provides a survival benefit is of interest, in their context, only if it is *heritable*. This is understandable, since a non-heritable mechanism may be of no significance in the broad sweep of evolution: it may vanish with the death of the individual that possessed it rather than remaining in the gene pool to serve as the basis of evolutionary development.

In general, a non-heritable action mechanism ordinarily represents a "dead-end" from the standpoint of evolution. It need not be a dead-end, however, for the individual organism that possesses it: that organism can owe its continued existence, and therefore its continued performance of that action, to the survival benefit provided by an action determined by a non-heritable mechanism. (Since virtually all the mechanisms for vegetative actions *are* genetically determined, and thus in principle could be passed on to offspring, the only non-heritable mechanisms would be those which confer an intra-lifespan benefit at the price of rendering the organism unable to reproduce successfully.)

In evolutionary biology, therefore, the form of natural selection that is relevant is *genetic* selection—the selection of a given genotype over the course of generations. As Dobzhansky states, "The essence of natural selection is the differential reproduction of the carriers of different hereditary endowments."[21]

A summary statement of genetic selection is given by Ayala:

> Natural selection is understood today in genetic and statistical terms as differential reproduction. Differential reproduction is a compound process, the elements of which are differential survival, differential mating success, and differential fecundity. Natural selection implies that some genes and genetic combinations are transmitted to the following generation on the average more frequently than their alternates. Such genetic units will become more common in every subsequent generation and their alternates less common. Natural selection is a statistical bias in the relative rate of reproduction of alternative genetic units.[22]

I would define "natural selection," in the sense of genetic selection, as: the process in which differences in the survival value of alternative heritable traits cause corresponding differences in the relative frequency of those traits in subsequent generations.

In contrast to genetic selection as described above, the differential perpetuation of alternative forms of action within the lifespan of the individual organism might be called "ontogenetic selection." This usage is drawn from the biological distinction between "ontogeny" and "phylogeny": "Ontogeny is the change in an individual from its formation until its death, while phylogeny is the change of organisms through a succession of generations."[23] (In order to conform to current biological usage, I will retain the term "genetic selection" rather than introduce the term "phylogenetic selection" to denote the differential perpetuation of alternate genetic units.)

The important point is that both genetic and ontogenetic selection give rise to teleological causation. The difference is only this: ontogenetic selection explains the *maintenance* in an individual organism of an intact mechanism for a given action; genetic selection explains the *origin*, in that particular organism, of that mechanism. Genetic selection is a statistical bias favoring the reproduction of the genotype causing the development of that mechanism.

For instance, in the case of the heartbeat we saw that the items carried in the bloodstream are responsible for the *maintenance* in a given individual of the heart. But in addition, the development of the heart in that individual animal can be explained teleologically by reference to its genetic endowment, which is in turn a result of the genetic selection of mechanisms that contribute to the survival of that kind of organism. Actually, I have already employed the principle of genetic selection, without identifying it as such, in explaining the existence of the auxin mechanism for phototropism. I explained the presence of the auxin mechanism in one generation in terms of a goal that mechanism achieved in the preceding generation. I asked the question: Why have plants with the auxin mechanism been able to reproduce with greater success than similar plants lacking this mechanism?— and answered in terms of the survival advantage provided by the phototropism effected by this mechanism.

By the same reasoning, it is possible to give a teleological explanation of the *origin* of the heart in a given animal. According to the principle of genetic selection, it is possible to explain the perpetuation in the population of the gene combinations that are responsible for the development of the heart in a given individual. And this explanation is *teleological* in that it refers to the contribution the heart has made to the survival and reproductive efforts of the parent animals. The chain of causation can be indicated by the following series of questions and answers: How is it that this animal has a heart?—because it possesses the gene combination for developing a heart; how is it that it possesses that gene combination?—because its parents reproduced; how is it that the parents were able to reproduce?—because (in part) they possessed functioning hearts. Or, looking at it negatively: this animal would not have a heart if it had not come into existence possessing the gene combination for developing it; it would not have come into existence possessing that gene combination if its parents had not been able to reproduce; its parents would not have been able to reproduce if they had not had functioning hearts.

In this manner, the present animal's development of a heart is ultimately explained by reference to the survival value of its parents' hearts. (Here, the survival benefit is both to the present animal and to its parents, since anything which is necessary to the survival of parents is *a fortiori* beneficial to their offspring: the offspring would never have come into existence if their parents had not survived to reproduce.)

It is important to stress that genetic selection explains the origin of mechanisms, such as the heart, only in the second and later generations. The original emergence in evolution of a novel feature, a novel modification of an existing feature, or a novel utilization of an existing feature is thought to be the result of chance-based processes, such as mutation and the genetic recombination resulting from sexual reproduction—it obviously cannot be attributed to selection nor explained teleologically. (Very frequently an existing feature which is already adapted to a given biological role happens, for non-teleological reasons, to be utilized by an organism in a new manner which proves to be adaptive in terms of a new selection pressure. In such cases of "preadaptation," the feature itself can be explained teleologically in terms of its old biological role, but its utilization in the new manner may still be regarded as the result of accidental, non-teleological factors.) Bock writes:

The origin of all features, be they adaptive or nonadaptive and the appearance of their later modifications lie outside the control of natural selection. The origin of these features—for example, the basis for the phenotypic variation to be acted upon by selection—lies under the control of a different set of evolutionary mechanisms and phenomena such as (a) mutations, recombinations of all sorts, gene flow, and other chance-based genetical processes that generate the genotypical variation underlying the phenotypical variation; (b) the nature of the pre-existing features of the ancestral group; (c) the geographical and ecological location of the ancestral group; and (d) the timing of events, such as which group is first to acquire a new feature and thus be able to exploit a new adaptive zone. . . . The common and significant property of all these factors is that they are chance-based (stochastic) with respect to the demands of selection, to future evolution of the feature, and to the future selection forces that will act upon the feature.[24]

As J. B. S. Haldane puts it:

To sum up, it would seem that natural selection is the main cause of evolutionary change in species as a whole. But the actual steps by which individuals come to differ from their parents are due to causes other than selection.[25]

For instance, this individual animal's development of a heart can be teleologically explained by reference to the survival advantage which having a heart conferred upon its ancestors. But each *new* step in the gradual phylogenetic evolution of the heart resulted from accidental, rather than teleological, factors. As a general rule, to whatever extent the features and activities of offspring are the *same* as those of their ancestors, genetic selection can play a role in explaining the offspring's possession of those features and performance of those activities. To whatever extent the features or activities of the offspring are *different* from those of their ancestors, genetic selection does not play a role in explaining those differences. The development by a given individual organism of a given adaptive mechanism can be explained by genetic selection—and hence teleologically—only if that mechanism was also possessed by its ancestors. A novel mechanism (or novel modification or utilization of an existing mechanism) even if it should prove beneficial to survival, is fortuitous in its first appearance.

Natural selection, whether genetic or ontogenetic, cannot anticipate the future. Adaptive mechanisms are selected by the survival benefit they have provided organisms in the past. Consequently, when environmental conditions change, a previously adaptive feature may become maladaptive, in which case natural selection begins to eliminate what it had previously favored. A classic example of this phenomenon is provided by H. B. D. Kettlewell's study of "industrial melanism" in a species of British moth. The results of that study are summarized by Gavin de Beer:

> Up to 1848 the British Peppered Moth existed in its typical grey form known as *Biston betularia*, which is remarkably well adapted to resemble the lichen on the bark of trees. From that date a dark melanic variety appeared, known as *carbonaria*, which is extremely conspicuous against the natural bark of the trees. It is controlled by a single dominant Mendelian gene and is slightly more vigorous than the normal grey type. Nevertheless, because of its conspicuous colour the *carbonaria* variety was constantly eliminated, and this variety only persisted in the populations of the Peppered Moth because the same mutation kept on occurring again and again.

(Note that this means that the presence of the dark pigment could not, at that time, be explained teleologically—it occurred only as a genetic accident.)

> The Industrial Revolution brought about a marked change in the environment, since the pollution of the air by increasing quantities of carbon dust killed the lichens on the tree and rendered their trunks and branches black. Under these conditions it is the *carbonaria* variety which is favored and the *betularia* penalized. This has

been proved by direct observation of the feeding of
birds, and by measurement of the survival rates of different forms in the different environments. The dark *car-
bonaria* form survives 17% less well in an unpolluted area
and 10% better in a polluted area. One hundred years
ago the dark variety of the Peppered Moth formed less
than 1% of the population: today in industrial areas it
forms 99%, and selection has made it more intensively
black than when it first appeared.[26]

Consider the case of a moth larva which grew up to be a
gray moth on a newly blackened tree. Although the gray coloring would no longer be adaptive, the moth developed it
anyway, and, let us say, was consequently eaten by a bird. This
illustrates the blindness of vegetative action: the *future* survival
significance of the goal has no causal influence on the development toward that goal. The process of development occurs
because of the beneficial effects that *past* instances of its goal
have provided the organism. (Thus, in this case, it would be
correct to say of this moth that it developed the gray coloring
in order to blend in with its environment, but that this action
failed to achieve that goal.) Normally, what has been beneficial in the past will continue to be so in the future; in those
statistically rare cases in which a changed environment revers-
es the survival value of a previously beneficial action, natural
selection takes its toll.

In addition, this example serves to separate the effects of
genetic and ontogenetic selection. Suppose that the *carbonar-
ia* moth's black color has to be sustained by an active process
involving the continual formation of new melanic pigment.
In this case, as soon as it would be the case that the moth
would have been eaten by a bird, were it not for its maintenance of the black color, we could say that its continued ability to form the melanic pigment is due to the survival benefit
which past acts of melanic pigment formation have provided—and thus that that action has been ontogenetically selected by its survival value. This would be sufficient to render its
subsequent performance of that action teleological—even
though no genetic selection would have yet occurred. Genetic

selection applies only after the moth has reproduced. At that point we can say that the gene combination for the pigment formation has been genetically selected for its survival value, and thus the origin of the mechanism for that action in the original moth's descendants would be explained teleologically.

SUMMARY: SELECTION AS THE BASIS OF GOAL - DIRECTED ACTION

Goal-directed action has the appearance of an impossible circle: the goal causing the very action which brings the goal into existence:

When analyzed more closely, the process is seen to be spiral, rather than circular: the goal that causes the action is an earlier instance of the goal that the present action will bring into existence:

$$\ldots \text{action}_1 \rightarrow \text{goal}_1 \rightarrow \text{action}_2 \rightarrow \text{goal}_2 \ldots$$

In this diagram, "action$_2$" represents a later performance of the same action as designated by "action$_1$." Likewise, "goal$_2$" refers to a later instance of the same goal as designated by "goal$_1$." (The subscripts "1" and "2" indicate any two successive occurrences of the action, not necessarily its very first and second occurrences.)

For instance, in the case of the heartbeat and its goal of circulating the blood, we have seen that the blood circulation produced by a given instance of the heartbeat is what makes possible the later instance of the heartbeat:

... heartbeat$_1$ → blood circulation$_1$ → heartbeat$_2$ → blood circulation$_2$ → ∴ ..

On the vegetative level, as on the purposeful level, teleological statements should not be understood as implying, in Nagel's phrase, that "goals or ends . . . are dynamic agents in their own realization."[27] In vegetative action it is a *prior* instance of the goal which is the dynamic agent in the action which realizes a later instance of that goal. Thus, it is possible to say that goals are dynamic agents in their own re-attainment. And, since the goal exerts its influence through the survival value it has provided, a better formulation would be: "The goal's survival value is a dynamic agent in its own re-attainment." A goal-directed action is a successful action repeated by virtue of its success.

The causal influence of the past goal is indirect, operating via its contribution to the enhanced survival of organisms "programmed" to perform the kind of action that has attained the goal. The connection of the action to the goal it attains, on the other hand, is direct: the action is a proximate cause of the goal. Therefore, the action causes the goal on a different level than the goal causes the action. This distinction may be represented in the diagram for goal-directed action as follows:

... action$_1$ → goal$_1$ → survival$_1$ → action$_2$ → goal$_2$ ...

It is important to stress, however, that we are referring to real, scientifically ascertainable, efficient causation in each case. The goal *causes* the organism's survival in the same sense that the action causes the goal; the survival of the organism is similarly an efficient cause of the organism's repetition of the successful kind of action. The difference between the ultimate cause and the proximate cause is one of immediacy, not a difference in the metaphysical status of the causation involved. The action, under appropriate conditions, brings about the goal; the goal, under appropriate conditions, brings about the organism's survival; the organism's survival, under appropriate conditions, brings about a recurrence of the action. The goal could not have been attained if the action had not been performed; the action could not have been performed if the organism had not survived; the organism could not have survived if earlier instances of the goal had not been attained.

On a naive, or pre-philosophical, level, goal-directed action is understood by analogy to, or as an extension of, the kind of causation exhibited in purposeful action. Accordingly, a "goal-directed action" is understood as a self-generated action caused by the value-significance of its goal. On the vegetative level, the action is not guided by any conscious awareness of the goal or of its value to the organism. Nevertheless, such actions are legitimately classed as "goal-directed" in that their ultimate cause is the survival benefits such actions have provided in the past (and will continue to provide in the future, barring a change in either the environment or the makeup of the organism). In light of this fact, a more advanced definition of the concept of "goal-directed action" on the vegetative level would be the following:

> A "goal-directed action" is a self-generated action caused by the survival value of past instances of its goal.

I use the term "goal" at the risk of appearing circular rather than encumber the definition with: "the state of affairs now regarded as its goal," or the like. It is understood here that the past instances of the goal referred to are those that were attained by earlier instances of the same kind of action.

In addition, the survival value of the goal need only be an ultimate or indirect cause of the re-occurrence of the goal-attaining action. The proximate causes of the action are its fuel and its mechanism (e.g., glucose and the mitochondrion in the case of respiration). However, the existence of this organized mechanism, its terms of operation (i.e., its subordination to specific directive mechanisms), and the organism's possession of the fuel for its operation will themselves be contingent upon the organism's survival, and thus derivatively contingent upon the survival value of the action in question. Hence, a more technical definition of "goal-directed action" would be:

> A "goal-directed action" is a self-generated action caused by a mechanism whose existence, fuel, and terms of operation result from the survival benefit gained from past instances of that action.

In this definition, the mechanism is said to be explained in that either one explains why the mechanism, once developed by the organism, continues to exist in that organism, and/or one explains the origin of that mechanism in the development of a given organism. The former is explanation via ontogenetic selection and the latter is explanation via genetic selection. In both cases, for the action to be goal-directed it must be true that the mechanism as described would not now exist but for the survival benefit provided by the kind of action it determines.

According to this conception, both adaptive vegetative action and purposeful action qualify as goal-directed: they are each caused by the value-significance of the kind of goals they attain. The issue that differentiates purposeful action from vegetative action is the involvement of consciousness. As I will argue in the next chapter, once we realize that consciousness itself is an adaptive mechanism and is accordingly to be explained in terms of its survival value, we can see that purposeful action satisfies the above definition of "goal-directed action."

VIII
OBJECTIONS I

In the preceding chapters I have argued that both vegetative and purposeful action exhibit teleological causation, and that a valid concept of "goal-directed action" wide enough to distinguish both from inanimate processes can be defined. In my analysis of purposefulness, I showed that an action qualifies as *purposeful*, when three conditions are met:

> 1. The action is performed by the agent (rather than being something which merely happens to the agent).
> 2. The action's goal has psychological value-significance to the agent (e.g., the agent *desires* the goal).
> 3. The action is caused by the goal's psychological value-significance to the agent.

I have argued that when any of these three features are absent, the action is not purposeful.

Our idea of teleological causation (or goal-directed action) however, is not so narrow as that of purposeful action. I have suggested that we consider an action teleological (or goal-directed) when three corresponding, but more generalized, conditions are met:

> 1. The action is self-generated.
> 2. The action's goal has a value-significance to the agent.
> 3. The action is caused by the goal's value-significance to the agent.

Accordingly, purposeful action is simply the most obvious way in which these conditions for teleological causation (or goal-directed action) are realized. In purposeful action the agent is *conscious* and consequently can experience value-significance in the form of a conscious desire and can direct its action on the basis of a conscious extrapolation of the future, based on past pleasurable and painful experiences.[1]

Vegetative action qualifies as teleological in a different manner, since there the involvement of consciousness is lacking. In the preceding chapter I argued that the natural selection of adaptive behavior implies that vegetative action does exhibit teleological causation, since it is true of vegetative action that:

> 1. The action is self-generated (rather than merely being something that happens to the organism).
> 2. The action's goal has a biological value-significance to the organism (i.e., the organism needs the goal to survive).
> 3. The action is caused by the goal's biological value-significance to the organism (as a result of the survival contribution the organism has derived from past instances of the goal).

Although the present instance of a vegetative action is not based on conscious extrapolation from the past (as it is in the case of purposeful action), vegetative action is still the result of a process of selection whose standard is the organism's survival needs. The factor common to purposeful and vegetative action is that the action is performed because it has had a value-significance to the agent. Both purposeful and vegetative actions are selected for performance on the basis of their past success. The difference is that in purposeful action the selection involves consciousness but in vegetative action it does not.

CAUSATION BY SIMILAR PAST GOALS: IS IT TRUE GOAL - CAUSATION?

This gives rise to a fundamental objection to the thesis that vegetative action is goal-directed. The objection is that since the thesis attributes causal efficacy not to the future goal of a given vegetative action, but to *past* instances of that goal, we are not here dealing with bona fide goal-directed action nor with genuinely teleological causation. Teleological terms

like "goal" and "end" specifically denote the future consequences of an action. But by the proposed analysis, the future
consequences of vegetative actions are not causal agents.
Therefore it is not the *goal* (i.e., the future state of affairs)
which causes or directs the action. We may grant, the objection holds, that a previous end state similar in nature to the
future goal is an indirect necessary condition of the action's
occurrence, but the previous end state is not the goal—the
goal is the future end state.

In short, the difficulty is that the object which causes the
action is not the action's goal, and the object which is the goal
does not cause the action. What causes a given vegetative
action is not *its* goal, but the goal of a similar previous action.

(It is hopeless to attempt to elude the objection by recast-
ing the thesis on an epistemological, rather than ontological,
level. It is not possible to bypass the question of teleological
causation by speaking instead of teleological explanation.
Causation is the necessary ontological basis of explanation;
explanation necessarily involves stating a cause. A shift to the
level of teleological explanation as an attempt to avoid
difficulties concerning teleological causation is thus question-
begging.)

The objection, however, "proves too much." For if the
objection leads us to regard vegetative action as non-teleological, it has the same result for the case of purposeful action. If
a goal-directed action implies goal-causation, and if only the
future end qualifies as the goal, then there is no such thing as
goal-directed action on either level—vegetative or conscious—because in neither case is there causation by the
future. Nothing in the universe qualifies as teleological, not
even man's deliberate, self-conscious, purposeful actions,
according to the standards set by the objection.

When, for instance, a student studies with the goal of passing an examination, the future state of affairs that he wishes
to bring about does not yet exist and hence cannot in any way
cause his actions. All that exists before the completion of his
action is his mental content—his desires, beliefs, expectations,
etc. This mental content, in turn, is not caused by the future

goal (which does not yet exist and might never exist). The mental content is caused by the student's extrapolations from past experiences with similar previous actions. In describing the student's action as goal-directed we imply (1) he believes his action may have a given effect, and (2) he desires the presence of those effects. His belief in the efficacy of study was derived in one or more of the following three ways: either by having himself studied in the past and having remembered its effects, or by having observed (or been informed of) the effects of study on others, or, never having studied for an examination himself and never having known of anyone else who had, he grasps what the effects of studying will be from his experience with other similar activities (e.g., memorizing, reading, and other forms of learning).

By whichever means he arrives at the belief that studying will improve his performance on the examination, that belief is based on the similarity existing between his present circumstances and *relevantly similar circumstances in the past.* The cause of his belief in the efficacy of studying is his extrapolation from past instances of studying (or past instances of related learning activities). And it is this belief, which is based on the experience with past instances of the goal, that is causally operative in his present purposeful study.

The same kind of past-to-future extrapolation could be traced in identifying the source of his *desire* for the effects he expects his studying to have. It is never the future goal which causes this desire; the future goal does not exist until the completion of the action (the goal must here be distinguished from the goal-object [see chapter 6], and if the action fails, the goal never exists.)

Purposeful human actions are purposeful because they are caused by the agent's desire for some expected effect of his action. But both that desire and that expectation are caused, not by the future goal, but by the agent's extrapolations from similar previous actions and effects. If the "backward reference" of the proposed analysis is not a solid enough basis for classifying vegetative actions as goal-directed, then by the same token the similar "backward reference" implicit in purposeful action is not grounds enough for classifying them as goal-directed either.

It might be argued, in rebuttal, that indeed the backward reference of purposeful action is not, in fact, sufficient to make such action qualify as goal-directed, but that it does qualify as goal-directed on other grounds. If so, it is incumbent upon the objector to offer a more plausible analysis of what constitutes the purposefulness of conscious action than that which has been presented here. It would be necessary either to maintain that in some way the desire and expectation operative in purposeful action are a response to a future goal after all (rather than to past instances of the goal), or to give some analysis of the purposefulness of conscious action which does not depend upon the factors of desire and expectation. The usual attempts in the latter direction (e.g., definitions of purposefulness stated in terms of behavioristic features, such as negative feedback) have already been criticized. These are generally so wide that they render virtually all natural processes teleological, and they inevitably fail to account for the distinction between a goal and a persistent result (e.g., they cannot account for the fact that the heart is said to beat in order to circulate the blood, but not in order to produce a thumping sound).

As I have noted earlier, it would of course be possible to accept these wider definitions of "teleology" and "goal-directed action." Such widened definitions, however, would not eliminate the distinctions upon which the proposed analysis is based. The issue is not linguistic, but factual. It would remain true that one class of the actions considered "goal-directed" in the widened sense would exhibit a mode of causation fundamentally similar to that of purposeful action, while another class would not. This fact would result in the coining of new terms to mark that distinction, and simultaneously the widened sense of "goal-directed action" would fall into disuse, owing to its tendency to blur this distinction (and due to the fact that nothing of significance follows from the fact that an action is "goal-directed" in the widened sense).

Putting the point in biological terms, there is a real and

important distinction between actions which have been selected in evolution for the attainment of certain advantageous
ends and those which have not, and this distinction cannot be
subordinated to less significant distinctions, such as that
between results which are persistent and those which are not.

It might be thought possible to analyze the goal-directedness in terms of just the first two requirements given in this
book—i.e., to define it as self-generated action leading to a
result which is beneficial to the agent. As already noted, however, this analysis is too weak to rule out the case of accidental
benefits. A man's conscious action may benefit him in an
unforeseen way—but in such cases it cannot be said that he
acted *purposefully* to achieve that benefit. To gain a benefit
through purposeful action and to gain a benefit through accident are two entirely different things.

Similarly, on the vegetative level there are self-generated
actions which accidentally bring benefits to the organism, but
which are clearly not teleological. For example, consider a
wild plant that produces flowers. The adaptive value (or, as I
would put it, the teleological function) of flowering is that
flowers attract insects which promote cross-pollination. But
these flowers may coincidentally please a human being who
consequently waters the plant. The water may be beneficial to
the plant, but it cannot be said that the wild plant flowered *in
order to* be watered by human beings. Here we have a vegetative action which is self-generated and provides the agent with
a survival benefit, but which is not goal-directed toward that
benefit precisely because the benefit is accidental: there is no
causal connection between a benefit of that kind (being
watered) and the action (flowering) which happened to bring
about that benefit. A recognition of the causal role played by
past instances of the goal is necessary to a proper understanding of goal-directed action.

Thus, my overall answer to the objection is that the same
chain of causation is exhibited in both vegetative and purposeful action. In the case of vegetative action we have:

$$\text{action}_1 \rightarrow \text{goal}_1 \rightarrow \text{survival} \rightarrow \text{action}_2 \rightarrow \text{goal}_2$$

where, as before, "action$_1$" indicates an earlier instance of the same type of action as denoted by "action$_2$," and "goal$_1$" indicates an earlier instance of the same type of end state as denoted by "goal$_2$."

Similarly, in purposeful action we have:

$$\text{action}_1 \to \text{goal}_1 \to \text{conscious satisfaction} \to \text{action}_2 \to \text{goal}_2$$

with the same meaning attached to the subscripts. For instance, on the vegetative level:

$$\text{heartbeat}_1 \to \text{blood circulation}_1 \to \text{survival} \to \text{heartbeat}_2 \to \text{blood circulation}_2$$

and on the conscious level, let us take the case of the dog being trained to sit on command:

$$\text{sitting on command}_1 \to \text{praise by master}_1 \to \text{pleasure} \to \text{sitting}_2 \to \text{praise}_2$$

In the first case, it is the survival benefit obtained by blood circulation$_1$ which in part causes the occurrence of heartbeat$_2$ leading to blood circulation$_2$. Accordingly, I maintain, heartbeat$_2$ is directed toward blood circulation$_2$ as its goal. In the second case, it is the pleasure the dog derives from praise$_1$ which in part causes the occurrence of sitting on command$_2$ which results in praise$_2$. Accordingly, sitting on command$_2$ is (in this case consciously) directed toward praise$_2$ as its goal. We say that the dog sits *in order to* be praised by its master precisely when it is the case that the dog has learned to associate that action under those circumstances with the reward—otherwise, we might think the dog's action of sitting when the command was given was merely a coincidence, rather than a purposeful execution of the command.

I conclude that the role played by past instances of the action in attaining past instances of the goal—the "backward reference"—is essential in both purposeful and vegetative

125

action. Consciously selected actions are purposeful, despite their "backward reference," and thus the "backward reference" of naturally selected vegetative actions cannot be an objection to considering them goal-directed.

PSYCHOLOGICAL AND BIOLOGICAL VALUE - SIGNIFICANCE

The second objection that must be met in defending the essential similarity of vegetative and purposeful action is the one raised in chapter 4: the objection that my use of the term "value-significance" is equivocal.

This objection holds that the biological "value-significance" of vegetative action is not equivalent to the psychological value-significance of purposeful action, and thus that vegetative action does not actually qualify as teleological or goal-directed. The objection is that the biological "value-significance" of vegetative action, lacking any emotional or affective side, is not really a value-significance at all. The biological "value-significance," it is objected, lacks the "felt impact" of purposeful action, and thus presents only a pale reflection of the emotional stake which men and animals have in the outcome of their purposeful actions. Accordingly, it is only a metaphor to say that the survival contribution made by the end states of vegetative actions have a value-significance to the organism.

In my discussion of this objection in chapter 5, I indicated several ways in which psychological value-significance and biological value-significance are related. I maintained that psychological value-significance develops out of the experiences of pleasure and pain. If, for example, an animal were born without the capacity to feel either pleasure or pain, it could not have any affective or conative experiences at all; such an animal would have no basis for desiring anything. The same is true in the case of man. If a man were congenitally unable to experience either physical pleasure or physical pain, he would be unable to *value* anything, or to desire anything.[2] Such a man would not engage in purposeful action, because there would be nothing for him to gain or lose by acting.

It might be suggested that an individual of this kind would still have his life to gain or lose—but this begs the question: how could such a man value his life and desire to live if nothing that he does can affect him emotionally? If such a man could not want to achieve or avoid anything in his life, he would be indifferent to whether he lives or dies.

To maintain that all psychological value-significance develops from the experiences of pleasure and pain is not to endorse psychological hedonism. As I understand that doctrine (and as the dictionary[3] defines it), psychological hedonism maintains that the goal or motive of all purposeful actions is the desire to experience pleasure and/or relief from pain. As a corollary, it holds that men are incapable of choosing a state of lower expected pleasure over a state of higher expected pleasure (counting the relief from pain as an increase in pleasure). Neither hedonist tenet is implied by my thesis.

Firstly, the goal or motive of a purposeful action is normally not the achievement of an internal state of pleasure, but the attainment of some object. The *object* of desire is generally not the pleasure which attends goal-attainment; the object is the attainment of the goal itself. The *cause* of the desire may be in part the pleasure which is associated with the goal, but the *object* of the desire is normally the goal, not the pleasure. Aristotle's statement carries the correct implication: "Desire is just appetition of what is pleasant."[4] In contrast, hedonism, as a psychological thesis, maintains: desire is just appetition of pleasure. If, for example, a man buys a car, normally his goal—the object of his desire—would be the possession and use of the car, not the pleasure taken in the possession and use of the car.

Secondly, the fact that the general phenomenon of desire develops from the general phenomenon of pleasure does not mean that every desire is simply the result of the agent's hedonistic calculation of what objects would be the most pleasurable to obtain. Having had the experience of valuing something emotionally, based on pleasure, man has the ability to invest non-pleasurable objects with value-significance; this is presumably what occurs when a man places a moral value on performing an action which he expects to decrease his total long-range pleasure. My thesis is that this kind of action is possible only because men have previously experienced the type of valuing which is based on pleasure and pain. (There is no evidence that animals, however, have this ability to act against the impetus of pleasure and pain, or to form desires that are anything other than the appetition of what is associated with pleasure or pain-reduction.) Psychologist Magda Arnold writes in *Emotion and Personality*:

> For the hedonist, pleasure is the basic motive in all actions. Reflection will show that such a statement makes an exception the general norm. An object or a function gives us pleasure because it suits us in some way. We want it because it is something good to have or to do. Our wanting becomes a motive when we endorse it and let it lead us to action. Wanting, as an emotion, always tends toward its object, an object that has been intuitively appraised as good in some way. To appraise pleasure itself as something good to have or something to aim for, apart from any particular object, requires reflection and even deliberation. To intend pleasure rather than the object that attracts us requires an act of rational choice. The animal is incapable of such abstract deliberations. He acts as his emotions urge him to act, and they follow the intuitive appraisal that some *thing* (whether perceived or imagined) is good to have. Only human beings, who can reflect upon their emotions and actions, and can initiate action even without emotion, can conceive pleasure in the abstract as a motive.[5]

Let us then focus our attention on animals, whose behavior is always motivated by the pleasure-pain mechanism (bearing in mind that the *object* of its desires and actions is the goal, not the pleasure which it associates with that goal).

All the conscious animals below man in the evolutionary scale act automatically toward that which they associate with pleasure and away from that which they associate with pain. (And, in the case of specifically human behavior, though pleasure and pain do not necessitate action, the pull toward the pleasurable and away from the painful is inherent in the very experience of pleasure and pain.)

On this basis, what can we say about the relation between psychological value-significance ultimately based on pleasure and biological value-significance based on survival—e.g., the relation between desire and survival needs? Are they only analogous, or is there a fundamental element common to both?

The answer is provided by looking at the faculty of consciousness from a *biological* perspective.

The faculty of consciousness in general and the pleasure-pain mechanism in particular are biological adaptations and should be understood in the same manner as every other such adaptation. The higher animals possess the faculty of consciousness because it has survival value.

This view of consciousness originated with Aristotle. Aristotle noted that since animals move around to obtain their food, they must have some faculty for locating that food, as well as for guiding their locomotion to avoid dangerous objects, predators, etc. To put it simply: mobile organisms need to be able to see where they are going. Aristotle, unfortunately, lacked a scientific explanation of how this need for consciousness could explain the presence in animals of a faculty to fulfill that need: he relied simply on the generalization that "nature does nothing in vain." Modern evolutionary biology can explain the present existence of the faculty of consciousness in terms of natural selection: consciousness has survival value. The expansion of consciousness meant the increase in the organism's range of perception, discrimination, and response, hence selection favored it.

The same functional explanation can be given for the pleasure-pain mechanism of consciousness. This mechanism exists because of its ability to provide the animal with survival benefits. Those survival benefits consist in the pleasure-pain mechanism's ability to motivate the animal to direct its conscious actions toward objects which aid its survival and prevent its death. R. A. Fisher notes that "the tastes of organisms, like their organs and faculties, must be regarded as the products of evolutionary change, governed by the relative advantage which such tastes may confer."[6]

It is readily apparent that the things which bring pleasure to an animal are the things which the animal's life requires; the things which bring pain are those which endanger its life. Illustrations of the empirically observable connection between pleasure and survival on the one hand and pain and death on the other were given in chapter 5—for instance, damage to any part of the animal's body causes it to feel pain. This connection cannot be just a fortunate accident. The nervous system's pleasure-pain mechanism is organized as it is because that organization has been adaptive to the survival of animals possessing it. It is not an accident, for instance, that bodily injury causes pain rather than pleasure (or nothing at all)—if such were not the case, the animal would not tend to withdraw from or avoid bodily injuries. Animals born without the capacity to experience bodily pain would stand little or no chance of surviving long enough to reproduce their kind. Similarly, animals so constituted that injurious actions brought pleasure would stand little chance of surviving long enough to repeat those actions. A dog so constituted as to find wounding itself pleasurable would bite itself until it bled to death, immediately eliminating from the breeding pool the gene(s) responsible for this wrong wiring of the pleasure-pain mechanism.

The stimuli which are inherently pleasurable to an animal are those associated in its natural environment with conditions required for its survival. Clearly, if an animal took no pleasure in eating, it would not be motivated to eat and would die. (It is interesting to note that the motivation provided by the desire to alleviate the pain of hunger would not suffice to keep an animal properly nourished; the animal would probably not live long enough to discover through trial and error those few instances that would alleviate its hunger, if it were not positively attracted by the pleasant scent and/or taste of those substances.)

The point is well stated by William James:

It is a well-known fact that pleasures are generally associated with beneficial, pains with detrimental, experiences. All the fundamental vital processes illustrate this law. Starvation, suffocation, privation of food, drink and sleep, work when exhausted, burns, wounds, inflammation, the effects of poison, are as disagreeable as filling the hungry stomach, enjoying rest and sleep after fatigue, exercise after rest, and a sound skin and unbroken bones at all times are pleasant. Mr. [Herbert] Spencer and others have suggested that these coincidences are due, not to any pre-established harmony, but to the mere action of natural selection which would certainly kill off in the long-run any breed of creatures to whom the fundamentally noxious experience seemed enjoyable. An animal that should take pleasure in a feeling of suffocation would, if that pleasure were efficacious enough to make him immerse his head in water, enjoy a longevity of four or five minutes. But if pleasures and pains have no efficacy, one does not see. . . . why the most noxious acts, such as burning, might not give thrills of delight, and the most necessary ones, such as breathing, cause agony.[7]

The correlation between the biological value of a substance and the pleasantness of its taste is often remarkable. Curt Richter[8] found that while normal rats avoid water containing calcium, apparently finding its taste unpleasant, rats rendered calcium-deficient by surgical removal of their parathyroid glands began to drink water containing the needed calcium, preferring it to their normal, pure drinking water.

There are, of course, notable exceptions to the general correlation of the pleasurable with the beneficial and the painful with the injurious. Arsenic is said to taste sweet; many human beings find the taste of alcoholic beverages agreeable; cats find tuna fish attractive, although it apparently sightly impairs their health. But none of these items—arsenic or alcohol in the case of man, tuna in the case of cats—are part of the natural environment to which the organism is adapted. Thus, natural selection has not had sufficient time to operate on these tastes in the new environment created by man, as James recognized:

> The exceptions to the law are, it is true, numerous, but relate to experiences that are either not vital or not universal. Drunkenness, for instance, which though noxious, is to many persons delightful, is a very exceptional experience. But, as the excellent physiologist [August] Fick remarks, if all rivers and springs ran alcohol instead of water, either all men would now be born to hate it or our nerves would have been selected so as to drink it with impunity.[9]

The adaptation of the pleasure-pain mechanism in man or animals is not flawless and unerring, but the overall connection—the fact that pleasure and pain are instruments of survival—is undeniable on both empirical and theoretical grounds. Empirically, the sheer number of cases in which the physical sensations of pleasure and pain do correlate with survival needs is astronomically greater than the number of non-correlating cases. Theoretically, one need only recognize that pleasure and pain exert an appreciable influence on behavior to realize that natural selection could not avoid operating to bring these responses in line with survival needs (assuming the pleasure-pain mechanism is genetically based).

Some ask, "How can you be sure of that? Is natural selection so omnipresent, so failproof?" But natural selection must not be reified; "natural selection" is not the name of a process which can occur or not occur. "Natural selection" means the law of causality applied to life: dead organisms don't act and don't reproduce; to go on acting, an organism has to act in the *right* way. Life is not a game: the losers cannot return to play again another day. Applied to the present case, this means: life-sustaining motivation gets maintained, life-threatening motivation gets wiped out. There is no question, then, of natural selection coming into operation or not doing so—anymore than one would wonder whether in a given case the law of cause and effect "came into operation." Natural selection is inescapable for living entities, by virtue of their being alive. If (1) pleasure and pain affect the initiation or direction of behavior, and (2) pleasure and pain contingencies, on the physical level, are due to the genetically based wiring plan of the nervous system, then pleasure and pain will necessarily be pulled into line with the pro-life and the anti-life, respectively.

Point (1)—the impact upon behavior of pleasure, pain, and the emotions deriving from them—is the very point that generated the objection we are here considering. The objection emphasizes that purposeful actions are caused by psychological value-significance. Evidently, the question becomes whether psychological value-significance is based upon anything heritable—as point (2) requires. (Let us leave aside specifically human action for the moment). Can it be that animal psychology is an unalterable given, causally insulated from the effects of natural selection? Surely one must grant that the springs of animal behavior lie in the makeup of the animal's nervous system and that the nervous system evolved in the same way that every other aspect of the animal evolved, subject to the same basic fact: what tends toward death is eliminated, what leads to life is sustained. To maintain that pleasure and pain contingencies are not governed by the inherited makeup of the species would be to render miraculous the uniformity among species of the pleasure-pain mechanism; why else would all men, for instance, feel pain when burned, jabbed with a pin, etc.?

I conclude, then, that it is natural selection which explains the correlation of pleasure and pain with survival needs. Specifically, it is genetic selection that explains why the things which bring animals pleasure and/or alleviate their pain are in general the very things which favor their survival. In evolution, the nervous system's pleasure-pain mechanism has become adapted to the survival of the animal. The pleasure-pain mechanism serves to motivate the animal's actions toward those things which its survival requires and away from those things which threaten its survival.

Likewise, it is ontogenetic selection which explains why the pleasure-pain mechanism of any given animal continues to exist and function throughout the animal's lifespan. An animal deprived of the guidance provided by this mechanism would come to a premature death, due to its inability to make vital discriminations and due to a general loss of motivation to take the actions its survival requires. Hence the continued operation of the pleasure-pain mechanism in a given individual animal is due to the survival benefits that past instances of its operation have provided that animal.

The general relation of physical pleasure and pain to survival is summarized by Rand:

> The pleasure-pain mechanism in the body of man—and in the bodies of all the living organisms that possess the faculty of consciousness—serves as an automatic guardian of the organism's life. The physical sensation of pleasure is a signal indicating that the organism is pursuing the *right* course of action. The physical sensation of pain is a warning signal of danger, indicating that the organism is pursuing the *wrong* course of action, that something is impairing the proper function of its body, which requires action to correct it. . . .
>
> The pleasure-pain mechanism of man's body is an automatic indicator of his body's welfare or injury, a barometer of its basic alternative, life or death.[10]

It might be objected that what is correlated with survival needs is only *physical* pleasures and pains, not more complicated emotional states such as human psychological satisfaction and frustration. We are indeed "programmed" to feel pain when our bodies are damaged, the objection holds, but the issue of what brings us emotional rewards or penalties of a more abstract sort is determined by such non-physical factors as our beliefs and values.

This objection has a point—for instance, one's literary tastes are certainly not biologically determined by one's inherited genetic makeup. My point is rather that one's more abstract desires and preferences are ultimately made possible by the basic alternative of physical pleasures and pains. In the formation of his beliefs and values, one has an element of volitional choice such that one is neither biologically nor environmentally determined to arrive at any given set of value-judgments. Man is not determined to desire what is in fact conducive to either his maximal pleasure or survival. But it remains true that the basis of *all* psychological value-significance is the biological alternative of life or death. The case for the basic similarity of vegetative action (based on survival needs) and purposeful action (based on desires) rests on the point that desire as a phenomenon has evolved because it has *biological* value. To see this, let us ask the question: Why did man's psyche—which admittedly gives him the power to act self-destructively—develop in evolution? The answer must be given in terms of the survival advantage such a psyche confers on man.

To put it in the crudest terms: man's brain does give him the power to desire objects which hinder his survival—but that's not why it's there. Clearly the faculty of consciousness in man and the higher animals is, on net balance, phenomenally adaptive, and this adaptiveness is the explanation of its evolutionary development. Thus, desire itself has a biological role; the psychological value-significance associated with desire is the psychological correlate of a more basic biological value-significance.

If we metaphorically personify the process of natural selection, we could say that the psychological states of pleasure and pain, desire and aversion, and satisfaction or frustration are simply Nature's means of having the alternative of life or death motivate the conscious actions of sentient organisms. The animal's emotional stake in its purposeful action is, then, not a primary, but rather derives from its biological stake in those actions. The psychological value-significance involved in purposeful action—far from being *sui generis*—is merely the form in which conscious organisms experience the fundamental biological value-significance of their actions. As Jonas puts it:

> Living things are creatures of need. Only living things
> have needs and act on needs. Need is based both on the

necessity for the continuous self-renewal of the organism
by the metabolic process, and on the organism's elemen-
tal urge thus precariously to continue itself. This basic
self-concern of all life. . . manifests itself on the level of
animality as appetite, fear, and all the rest of the emotions.[11]

The fundamental alternative for purposeful action—as for
vegetative action—is life vs. death; the alternative of pleasure
vs. pain (and satisfaction vs. frustration of desire) exists only
as a means to the maintenance of life in the face of the possibility of death.

Psychological value-significance and biological value-significance, then, are not *identical*, but they are fundamentally
related. In the light of the preceding discussion, we may say
that psychological value-significance is biological value-significance plus consciousness. This distinction between the two is
retained in my general thesis: what vegetative and conscious
action have in common is *goal-directedness*, not purposefulness.
Purposefulness is the specific type of goal-directedness manifested in conscious action. A purpose is a goal that is consciously anticipated by the organism and whose value-significance is consciously experienced in the form of a desire for
the goal.

By recognizing the biological function of consciousness,
we can see that purposeful action satisfies the most fundamental definition of "goal-directed action":

> A "goal-directed action" is a self-generated action caused
> by a mechanism whose existence, organization, fuel, and
> terms of operation result from the survival benefit that
> past instances of the goal have provided the organism in
> similar previous circumstances.

Taking the case of purposeful, conscious action, we have:

> A purposeful action is a self-generated action caused by a
> mechanism [i.e., consciousness] whose existence, organi-
> zation, fuel, and terms of operation result from the sur-
> vival benefit that past instances of the goal [i.e., past

adaptive purposes] have provided the organism in similar previous circumstances.

ALLEGED COUNTEREXAMPLES

Another general class of objections concerns the scope of the proposed analysis of teleological causation. Does the analysis include all those processes and only those processes which are commonly regarded as being teleological? Are some manifestly non-teleological processes rendered teleological by the proposed analysis? Are some relevant vegetative actions rendered non-teleological? How does the analysis handle cases that are controversial—i.e., cases of behavior which are neither clearly teleological nor clearly non-teleological? What is the relationship between the class of actions qualifying as teleological by this analysis and the class of living actions?

First, is the proposed analysis too broad? Does it include as teleological some types of process which exhibit no apparent purposefulness, and for which teleological explanations are generally regarded as inappropriate?

It is easy to see that the proposed analysis is at least not so broad as to render teleological concepts vacuous; the analysis clearly precludes many inanimate processes from being considered teleological. The process of a rock rolling to the bottom of a hill, for example, fails to satisfy any of the requirements for teleological causation: (1) the process is not a self-generated action (it is a passive response to an external factor: the earth's gravitational field), (2) the final state has no value-significance to the rock (whether or not the rock reaches the bottom of the hill makes no difference to the rock,[12] and (3) the process is not caused by any value-significance provided by past instances of the end state. The third point follows from the fact that no value-significance is involved and that the process is not a self-generated action—i.e., the third point is a consequence of the first two.

Similarly, to use Aristotle's example,[13] it cannot be said that rainfall occurs *in order to* help the crops grow: (1) the rainfall is not a self-generated action, (2) although the rainfall does have a value-significance to the crops, we are concerned only with its value-significance to the agent (see chapter 5) and the rainfall has no value-significance to the clouds, and (3) the benefit which the crops derive from the rainfall is not a necessary condition for the re-occurrence of the rain-fall—even if no crops had ever benefited by rainfall, the rain would still fall—as it did during the billions of years before such vegetation existed.

RAINFALL AND THE WATER CYCLE

Such cases as rocks rolling downhill and rainfall helping crop growth are clearly non-teleological according to the pro-posed analysis, since they satisfy none of the conditions pre-sented. It might be objected, however, that other very similar kinds of inanimate processes are rendered teleological by the analysis. What of rainfall itself, for instance? Rainfall is the product of a cycle of events in which earlier instances of end states are necessary conditions for the occurrence of later instances of the same kind of end state.

The proximate cause of the rainfall is the water vapor composing the clouds; the cause of the presence of water vapor in clouds (i.e., of the existence of clouds) is the water evaporated from lakes, rivers, and seas; the continued exis-tence of lakes, rivers, and seas is contingent upon their replenishment by rainfall. Thus present instances of rainfall have as an indirect necessary condition the contribution to the cycle made by past rainfalls. If the water evaporated from the earth in the past had been dispersed into outer space instead of returning to the earth in rainfall, there would be no source of water vapor for the present rainfall. Hence rain-fall, the objection holds, exhibits goal-causation by the pro-posed analysis since it is an action whose continued occur-rence is caused by the kind of goal it reaches.

It must be conceded that if the proposed analysis implies that rainfall is teleological, then the analysis has failed. Few processes could be farther removed from purposeful action—the paradigm of teleological causation—than rainfall.

Rainfall, however, cannot serve as a counterexample to the proposed analysis since it fails to satisfy two of its conditions: the rainfall is not a self-generated action and it has no value-significance to the "agent." I put "agent" in quotes because it is difficult even to isolate which part of the meteorological system is the agent of the rainfall. Unlike living organisms, clouds are not organized entities, but diffuse aggregations of relatively independent particles; this in itself makes it difficult to identify what would count as the "self" for a determination of whether the process of rainfall is self-generated. It might be best to speak of a cloud system as the agent in rainfall. Difficulties still exist, however, since the rainfall is the fall of the very water which composes the cloud system. Raining is not an action performed by clouds, it is the progressive transformation of the clouds into rain. Clouds do not *produce* rain—they turn into rain.

> When air reaches the dew point, some of the water vapor in the air condenses into tiny particles of water, so fine that they might be called *water dust.* This water dust is known as clouds or fog, according to whether it is high in the air or near the surface of the earth. A still greater cooling of the air will cause the tiny cloud particles to unite into drops so large and heavy that they fall.[14]

In rainfall, then, we do not have a case of an entity performing a self-generated action. There is no internal supply of energy, no action mechanism, and no directive mechanisms to channel the utilization of energy. Rainfall is not an action performed by clouds, but rather something that happens to water molecules that have been lifted from the earth's surface by solar energy. For this reason alone, rainfall cannot be considered teleological.

In addition, no *value-significance* can be ascribed to the results of the rainfall. The rainfall certainly does not benefit the clouds—it is the destruction of the clouds. From the standpoint of the water which composes the clouds, there can be no value-significance assigned to any stage of the cycle, because the water does not need to act in order to remain in existence and therefore has no needs.

Rainfall, although part of a cyclical process of events, is not teleological by the proposed analysis because it is not a self-generated action and because it has no value-significance to the "agent" (assuming one could even identify an agent in this case).

THE PENDULUM AND SIMILAR MECHANISMS

Another proposed counterexample might be the action of a pendulum. Each downward swing of the pendulum has as a necessary condition, *ceteris paribus*, the potential energy gained by its preceding upward swing. Assuming no external forces acting to raise the pendulum's bob, the only reason the pendulum is able to swing downward is that past instances of swinging downward gave it the momentum to pass beyond the equilibrium position and reach a height from which it will fall again. Hence the manifestly non-teleological behavior of a pendulum seems to satisfy the requirement for teleological causation: past instances of the goal (moving downwards) are an indirect cause of the repetition of the action by which that goal was achieved.

Here again, the objection is based on the erroneous assumption that cyclical causation is all that is required for a process to qualify as teleological by the analysis. As in the preceding case, the pendulum's behavior actually fails to qualify as teleological because it is not self-generated and because its end state, lacking any value-significance, cannot be described as a *goal.*

The fall of the pendulum bob is no more self-generated than the fall of a rock, the fall of rain, or the fall of a man who trips. It is true that the pendulum comes to a position from which it can fall as a result of the momentum gained in its own prior downward swing, but this does not serve to make the motion self-generated. The total energy for all of the swings is supplied by whatever external force raised the pendulum above its equilibrium position in the first place. The potential energy of the bob when it is above the equilibrium position is not energy stored within the pendulum itself, but pertains to the relationship existing between the pendulum and the earth's gravitational field: if the pull of the earth were eliminated, the pendulum's potential energy would vanish. The bob is quite obviously not self-moving, but is rather pushed and pulled about by the external force of gravity.

What if the pendulum is part of a clock such that the energy for its action is supplied not by an external factor, but by the clock's own mainspring? Is the clock's action self-generated? No. As explained in chapter 4, the release of internally stored energy is a necessary but not sufficient condition for a process to qualify as a self-generated action. The release of a spring is not a self-generated action firstly due to the fact that the spring's potential energy is not available for alternative actions, but only for the single action of expansion. The energy is primarily associated with the process of expansion, not with the "self"—i.e., the spring or the clock. To repeat a statement from the fourth chapter: There is no directive mechanism mediating between the energy supply and the expansion, as manifested by the fact that one could not speak of the spring as being "programmed" to expand (whereas, one could speak of a plant as being programmed to turn its leaves toward the sun).

It might be replied that in living organisms the ultimate source of energy is the binding energy stored within the ATP molecule, and that the exergonic conversion of ATP to ADP is nothing but the chemical equivalent of the expansion of the spring. This is quite true, but the conversion of ATP to ADP is not *per se* a self-generated action. Although the conversion of ATP to ADP is the ultimate *means* by which an organism performs self-generated actions, that chemical reaction in itself is not self-generated: "self-generation" refers to the manner in which the energy released in reactions of that type is controlled and utilized by the organism. In the pendulum clock, on the other hand, there is a direct one-to-one hookup of the spring to the pendulum. The energy stored in the spring cannot be utilized in alternative actions and is not integral to the clock as a whole. The clock can be disassembled, the pendulum separated from the spring, then reassembled, with no loss of function. In a living organism, the power is not supplied by a separable "motor," but is possessed by each of the parts themselves, is in fact part of the *structure* of these parts, none of which can maintain their energy outside of the organism (see chapter 4).

Another factor serving to differentiate the clock's behavior from self-generated action is the ultimate source of the energy for the action. Neither the clock nor any living organism is able to *create* energy, but there is still an important difference in the way clocks and living organisms obtain their energy supplies. The clock must be wound up by man; each living organism, in contrast, is able actively to appropriate energy supplies from the natural environment. In the case of living organisms, part of the energy supply is utilized in the process of obtaining new energy supplies, and this fact is included in our idea that living organisms can be actively self-moving in a way that a spring-wound clock cannot. As opposed to the behavior of the spring-wound clock, the self-generated actions of living organisms are not cases of "running down." As Lehninger states:

> It is the capacity to extract energy from its surroundings and to use this energy in an orderly manner that distinguishes the living human organism from the few dollars' . . . worth of common chemical elements of which it is composed.[15]

The spring-wound clock lacks the ability to extract energy from its surroundings. Of course, this particular objection could be countered by constructing a pendulum clock that is powered by a photocell which would transform the energy of sunlight into the mechanical energy for the pendulum's swings. Such a clock, however, is already quite removed from the original simple pendulum which was to serve as the counterexample, and the other objections against considering its behavior self-generated would remain in force.

Even if the motion of the pendulum in a clock were considered self-generated, the clock's behavior would not qualify as teleological, since there is no value-significance involved. Whether the pendulum swings or not makes no difference to the clock: the clock cannot be said to *need* the pendulum's swings, the swings are not beneficial to the clock, the failure to swing does not harm the clock. Unlike a living organism, the clock's existence is not conditional upon its motions. The swinging of the pendulum may indeed be *our* goal, but it is not the clock's goal. The clock, having no needs, can have no goals.

PSEUDO - NEEDS: AVOIDING DESTRUCTION VS. GAINING VALUES

What if the clock were equipped with an internal bomb arranged in such a manner that the failure of the clock's pendulum to swing would cause the bomb to detonate? Would not the clock's behavior in this case involve a value-significance, since only by continuing to function can it prevent its own destruction? No.

First, there is something essentially arbitrary about the "threat" posed by the bomb. The placement of the bomb in the clock and the bomb's specific terms of detonation are due only to the arbitrary choice of the human designer. The presence of the bomb is in no way required by the clock's function, nor is there any independent reason why the bomb is arranged to detonate when the clock's pendulum stops—it might just as well be arranged to detonate when the pendulum moves as when it stops moving. The behavior is *appropriate* not in terms of the clock, but in terms of the goal of its human designer who decided to wire the bomb into the clock to explode under the conditions he desired.

Second, and more importantly, a distinction must be made between the acquisition of benefits and the avoidance of harm. The concept of "harm" is dependent upon the prior concept of "benefit": something is harmful to the extent that it interferes with the acquisition of a benefit.

For instance, damage to the limb of an animal is harmful to the animal because it impairs the animal's ability to engage in the acquisition of positive benefits—e.g., to acquire food. In its role as a concept denoting value-significance, "harm" means more than just damage or destruction: it carries the implication that the damage or destruction would block or interfere with the agent's actions to obtain positive benefits. If an entity is not capable of engaging in actions to acquire positive benefits, it is not possible to describe damaging or destructive factors as *harmful* to that entity, if by using the term "harmful" we are attempting to ascribe value-significance to the actions by which such damage or destruction is avoided. The avoidance of damage or destruction can have value-significance only when there are some other actions to acquire positive benefits which that damage or destruction would impair. Thus, to argue that the clock's avoidance of the explosion of the bomb has value-significance, prior to showing that the clock has any positive goals, is to beg the question.

It might be objected that the actions of living organisms are themselves ultimately concerned with the avoidance of harm, rather than the achievement of positive benefits, and thus that if the analysis of "harm" given above is correct it precludes assigning any value-significance to the actions of living organisms as well. Actions are to be judged beneficial to an organism when they serve the organism's life, but the maintenance of life, this objection would hold, is ultimately just the prevention of the organism's disintegration. Food is beneficial to an organism, for example, in that it provides energy for the maintenance of the organism's structure against the tendency of physical systems to lose organization (i.e., for their entropy to increase). The statement of Lehninger quoted earlier supports this interpretation:

> A living cell is inherently an unstable and improbable organization; it maintains the beautifully complex and specific orderliness of its fragile structure only by the constant use of energy. When the supply of energy is cut off, the complex structure of the cell tends to degrade to a random and disorganized state.[16]

Doesn't this mean that life is the sum of the actions the organism undertakes to prevent a loss—the loss of organization? And isn't it true, therefore, that in the last analysis both the clock and the organism are acting to avoid a loss?

In answer to this objection it should first be noted that internal disintegration, due to lack of energy for self-maintenance, is quite a different thing from destruction by the action of an external factor, such as a bomb (or a predator). Although the bomb may be located physically within the clock's structure, it is nonetheless an *extrinsic* threat: The existence of the bomb is not required by the clock's functioning. The bomb could be removed or defused without in any way impairing the clock's ability to function.

The possibility of internal disintegration, however, is *intrinsic* to the functional nature of each living organism. Lehninger speaks of the cell as "inherently" unstable; the clock is not inherently or intrinsically faced with the threat of destruction—it is faced with this threat only by virtue of the arbitrary and revocable choice of its designer. In contrast, the possibility of disintegration (death) is necessitated by the organism's nature; inaction causes the internal disintegration of an organism not as a result of our choice, but by necessity. The actual counterpart in the case of the clock to the internal disintegration of the living organism is the gradual disintegration of the clock's structure, over centuries, due to rust, metal fatigue, etc. But then the clock's behavior does nothing to counteract this internal disintegration (its behavior in fact contributes to the disintegration by causing wear and tear on the clock's moving parts). Hence, the clock's behavior may be distinguished from the actions of living organisms first of all by the kind of "threat" faced by each: an intrinsic threat must be distinguished from an extrinsic one.

Secondly, to meet the objection head-on, it is not true that in feeding, growing, reproducing, etc., the living organism is simply avoiding a loss, in the manner of the clock avoiding the bomb's explosion. It is true that if the organism does *not* perform its vital functions it will die, but this does not imply that the vital functions are not positive achievements in their own right. By analogy, if a student does not pass a course, he will fail—but this does not imply that passing the course is nothing but the avoidance of failure: in order to pass the course, the student must obtain knowledge, which is no less a positive achievement for being undertaken in the face of the possibility of failure.

The difference between the acquisition of benefits and the avoidance of harm is this: *in avoiding harm the organism suffers a loss,* although less of a loss than it would suffer if it did not act.

For instance, if a zebra escapes from a pursuing lion, the zebra suffers a loss of energy—the energy required for the escape. The zebra is less well equipped for survival after its escape than it would have been had the lion never pursued it. On the other hand, using up that energy to escape the lion is preferable to being killed by the lion. Thus the zebra's avoidance action causes it to suffer a loss (but a smaller loss than it would have suffered by not acting). In acquiring a benefit, however, performing the action renders the organism better equipped to survive than it would be if it had not acted at all. When a zebra eats grass, the action is positively beneficial in that the grass's contribution to the zebra's survival exceeds the loss of resources involved in perceiving, approaching, eating, and digesting the grass.

Although both the zebra's action of eating and the clock's behavior make the continued functioning of each possible, they do so in fundamentally different ways: the action of the zebra not only prevents internal disintegration, but also provides the fuel for future actions; the behavior of the clock makes its future functioning possible merely by preventing interference from an external factor.

The avoidance of damage or destruction can have *value-significance* only where there are some *other* actions to acquire positive benefits which that damage or destruction would prevent. In the case of the clock with the internal bomb, there are no actions providing positive benefits, which means that its existence is not conditional in the required sense. Hence the clock's behavior has no value-significance and consequently cannot be described as goal-directed according to the proposed analysis.[17]

Life and goal-directedness are intimately related. For consider the three requirements of goal-directedness: self-generation, value-significance, and goal-causation. Each implies the others, and all are a consequence of the essential nature of life: conditionality. Self-generation means there is an internal store of energy. This store must be replenished—hence the *need* to obtain energy—hence the phenomenon of value-significance. What underlies goal-causation? The fact that only valuable actions get repeated. Why do only valuable actions get repeated? Because the value here is *survival* value, and to repeat the action, the agent must survive.

The preceding discussion may prompt the question of whether a machine could be built whose existence *would* be conditional upon a process of positively self-sustaining action. Is it possible to build a self-maintaining machine which can self-generatedly act to gain materials from the environment from which it can process to extract energy and repair its parts? Doubtless we will eventually be able to construct such a device—which is just to say that eventually we will be able to create a living organism. In conceiving of an artifact capable of goal-directedness, we are conceiving of a living organism. Since I am not a vitalist, I do believe that it is in principle possible to create life from non-living materials. My point is only that we do not even move closer to satisfying the criteria for goal-directedness (or for life) by merely placing a bomb in a spring-wound clock.

Hypothetical artifacts which incorporate many of the features of living organisms constitute borderline cases that, as such, cannot count either for or against my thesis. It may indeed be difficult to determine any exact point at which the actions of a self-sustaining machine/organism qualify as goal-directed, just as it is difficult to determine any exact instant at which day ends and night begins—there being a certain borderline area in each case. But this no more invalidates the distinction between goal-directed and purely mechanical processes than it does the distinction between day and night.

FIRE

A proposed counterexample drawn from the natural, rather than the man-made, realm is that of a flame. It might appear that a flame satisfies all the requirements of goal-directedness. It appears to be a self-generated action, its existence is conditional upon its actions, and its actions seem to secure it positive benefits: fuel and oxygen for continued combustion. Does not, then, the flame's action qualify as goal-directed by the proposed analysis?

But on more careful inspection, we can see that the flame does not in fact satisfy the proposed analysis of goal-directedness.

First, in order to be able to speak of an action's being self-generated, there must be a "self"—i.e., a physical entity with a defined structure and a definite boundary isolating it from the environment. A flame is not such a thing. A flame is not an entity at all, but a *process*. A flame is in the same category as a waterfall—neither are entities, in the primary sense of that term. The *entity*, in the case of the flame, is the object which is burning—e.g., a piece of wood. But the object which is burning is not being sustained or benefited by the process of combustion—the flame is *destroying* the object.

It has been objected that if a flame is not an entity, then neither is a living organism. The grounds offered for this objection are that, just as in a flame, the matter composing a living organism is in a continuous state of turnover, the molecules originally composing the organism being continually replaced over a period of weeks, months, or years, by new materials taken in from the environment.

There are a number of ways of answering this objection. First, it does not seem to be the case that all of the molecules composing all organisms are in fact replaced during their lifespans, nor that all the molecules within a given cell are replaced during the lifespan of that cell. In the single cell, the DNA chain apparently remains unaltered throughout the cell's lifespan.[18] And in the case of animals, the cells composing the nervous system are not replaced during the animal's lifespan. Since DNA and the nervous system are the ultimate "control centers" of the cell and the animal, respectively, it is significant that they are not subject to the general turnover.

Second, aside from the exceptions to the replacement process, the replacement of cells in the body and of molecules within the cell is not similar to the turnover of the particles composing the burning vapors of the flame. The organic replacements differ both in rate and in directiveness from that occurring within the flame. With regard to rate, there is an enormous difference between the split-second turnover of material in the flame and the gradual replacement of the matter composing an organism. This is a difference in degree, true, but most differences in kind are, in a wider context, reducible to differences in degree—e.g., the difference between night and day is a difference in the degree of ambient light.[19] Even a stone is eventually worn away by the atmosphere over millions of years, but a stone is clearly an *object* in a way that a flame is not.

Also the replacement of matter in an organism is systematically controlled by stable physical mechanisms which ensure that the new matter will take on the structure of the old. The new protein and enzymes of a living cell are *synthesized* in a specifically controlled series of reactions guided ultimately by the "information" stored in the DNA. This is quite a different thing from the turnover of material in the flame.

Third, the obvious common-sense differences in the status of organisms and flames remain. One cannot pick up a flame (as apart from the object which is burning); one cannot hold the flame in one's hands. The instant the flame "fails" in its allegedly self-sustaining action, it vanishes—it leaves no corpse to disintegrate. It is simply perverse to maintain that flames and living organisms belong in the same metaphysical category. The concept of "entity," which is the base of all our knowledge, is self-evident—and organisms are entities par excellence. The difference between a flame, as a process, and an organism, such as a tree or a man, is given perceptually, and no argument could undermine it.

Furthermore, the process of combustion is not self-generated. It is true that there is a release of internally stored energy in combustion, but: (a) because a flame is not an entity, it is impossible to speak of a "self" which possesses the energy (as apart from the "self" constituted by the object which is burning), and (b) there are no directive mechanisms for channeling the energy in alternative directions, there is no independence of input and output, no triggering.

It is true that on the vegetative level the actions I regard as goal-directed are strictly necessitated by the mechanical causes operating. A plant, for example, has no choice as to whether or not it will perform photosynthesis—when the mechanism is activated, photosynthesis occurs, when not, not. But in the case of the plant, as opposed to the flame, there is no one physically necessary mechanism. The presence in an organism of, e.g., the auxin mechanism or the heart, is not deducible from a physical law of nature. And the terms of that mechanism's operation—the link between stimulus and response—are variable. No one "program" for the input-output relations is physically necessary (although each variant is itself necessitated by antecedent causes).

What underlies the phenomenon of self-generated action in this respect is the fact that *alternative genotypes are physically possible.* But this variation is missing in the case of the flame. Thus, aside from all the other objections to considering the flame's action as goal-directed, the absence of any variable "programming" renders the action of the flame purely mechanical rather than teleological.

Alternative "programs" for actions, rooted in the varying DNA codes of organisms, provide the only way in which a mechanically determined process can qualify as teleological, since only in that case can we say that although the action was necessitated by the given mechanism, that mechanism and its terms of operation were *selected*—and they were selected because they lead to an action which in the given type of environment is beneficial.

"EXTERNAL" TELEOLOGY

Ayala, following the terminology of T. A. Goudge,[20] has suggested that teleological concepts are applicable to presently existing artifacts, although in a different sense than that in which they are applied to living organisms. Artifacts, in Ayala's view, exhibit "external teleology," as opposed to the "internal teleology" of living organisms. The teleology ascribed to artifacts is said to be "external" in that although the artifact has no goals of its own, it embodies the goals of its human designers.

> The end-directedness of living organisms and their features may be said to be "internal" teleology, while that of man-made tools and servo-mechanisms may be called "external teleology." . . . Internal teleological systems are accounted for by natural selection which is a strictly mechanistic process. External teleological mechanism[s] are products of the human mind, or more generally, are the result of purposeful activity consciously intending specified ends.[21]

To avoid confusion, it should be noted that "external teleology," as Ayala uses the term, would in fact apply to *all* man-made objects that serve a human purpose, not merely to those special kinds of devices, such as target-seeking torpedoes, which mimic some of the superficial features of living action (e.g., plasticity or negative feedback). A house or a painting, for instance, is fully as teleological, in Ayala's sense of "external teleology," as a target-seeking torpedo. The house and the painting are held to exhibit "external teleology" in that their existence and their specific properties are explainable in terms of the conscious goals of the human beings that created them.

The term "external teleology," however, is inappropriate: it widens the concept of "teleology" in such a way as to obscure its essential meaning. Basically, to say that a feature of an entity is teleological is to say that the feature benefits *that entity*, and that the benefit is the cause of the feature's existence; a feature is teleological if it is present because it is good *for the entity which possesses it.* "External teleology," however, would denote the fact that the features of some entities are explainable only in terms of their value to *other* entities. Thus, "internal" and "external" teleology do not make a distinction between two different types of teleological objects, but between two different senses of "teleology" itself.

To say that a hammer exhibits "external teleology" because its features are explainable in terms of human goals is like saying that a book exhibits "external knowledge." Although in an informal context we do refer to books as "containing knowledge," in the strict sense knowledge resides only in the minds of men, not in the symbols by means of which men record their knowledge. An epistemologist who proposed a distinction between "internal" and "external" knowledge along these lines would be obscuring the fact that the "knowledge" possessed by books is actually only a potential for causing knowledge to exist in the mind of a reader. Likewise, the "purpose" embodied in a hammer is only the potential for serving a purpose held in the mind of a human being.

The terms "internal" and "external," by their linguistic form, give the appearance of marking a distinction between items on the same level of analysis, whereas in fact "external teleology" is a derivative of "internal teleology," since the "goals" embodied in artifacts are caused by, and understood in relation to, the independently existing goals of man (i.e., "internal teleology").

It is thus preferable to restrict the term "teleology" to the single sense which Ayala and Goudge call "internal teleology," and to denote the products of human purposeful action (i.e., "external teleology") by the familiar terms "artifact," "device," "utilitarian object," etc.

IX

OBJECTIONS II

IS REPRODUCTIVE FITNESS, NOT SURVIVAL, SELECTION'S GOAL?

A question postponed from earlier discussion concerns the ultimate goal of the action of living organisms. Is the ultimate goal survival or reproduction? And is it the maintenance of the individual or of some group of which the individual is a member? Since teleology is based on natural selection, the question of the ultimate goal of teleological action is the question of the ultimate basis or standard of natural selection: what is it that selection favors—the survival of the individual organism? the reproduction of the individual? the continued existence of the group?

As explained in chapter 7, there are two kinds of natural selection relevant to teleology: "ontogenetic" selection and genetic selection. "Ontogenetic" selection explains the maintenance of some feature or mechanism in the makeup of an organism in terms of the contribution that feature or mechanism makes to the survival of that particular individual. For instance, my heart exists and functions today only because its existence and functioning throughout my life up to this time has enabled me to survive. Had I not possessed a heart, or had it not functioned properly earlier in my life, I would have died and my heart would have disintegrated. Thus, the continued presence of a beating heart in my body is conditional upon (among other things) the ability of my heart to keep me alive. Ontogenetic selection, then, is clearly based upon the survival of the individual—it exists whether the individual is able to reproduce or not, and whether the individual is part of a group or not.

Ontogenetic selection alone, without reference to genetic

selection, is sufficient to justify teleological explanation of the structures and actions of organisms subjected to it. My present heartbeat can be explained teleologically in that it has as a necessary condition the survival contribution I have derived from its past occurrences. My heartbeat occurs because it has made possible my continued survival; the continuation of my heartbeat is contingent upon its continued ability to sustain my life.

A deeper level, as it were, of teleology is based upon genetic selection. While ontogenetic selection explains the *maintenance* within a given individual of a feature or mechanism once acquired, genetic selection explains the origin within a given individual of a feature or mechanism (except for its evolutionary emergence in the very first organism(s) in which the feature or mechanism appeared). The ontogenetic development of any feature or mechanism is the product of the genetic "programming" set in the individual at the zygote stage. The existence of the genes which make up the zygote's genotype are explained in terms of genetic selection. Thus by reference to genetic selection it is possible to explain why I developed a heart in the first place, not merely why my heart remains functional once developed.

It is genetic, not ontogenetic, selection which is of concern to evolutionary biologists, because it is only genetic selection which explains the process of evolutionary development.

If an organism could be immortal, the perpetuation of the genetic units which conferred immortality upon it would be guaranteed without reproduction. Short of the impossible state of individual immortality, the perpetuation of a given gene allele or gene complex requires reproduction. Contemporary biologists, consequently, tend to look upon genetic selection as differential reproduction, and include survival as one requirement of reproduction (since a dead organism cannot reproduce). It is equally possible, however, to look at the same phenomenon from the other direction and describe genetic selection as differential survival, including the individual's genesis as a requirement of its present existence. If a genetic unit increases the probability of reproductive success, then *ipso facto* it increases the probability of the offspring's creation and development. Clearly the goal of "reproductive success" is not attained if the offspring cannot mature and survive. A gene allele which increases mating activity but causes all the progeny to be stillborn will have a reproductive fitness of zero and will rapidly be eliminated from the population. "Reproductive success" is understood as success in producing *viable* (and fertile) descendants. If this is so, then it is impossible to distinguish benefits to reproduction from benefits to the lives of the descendants.

The equivalence of the two ultimate goals—survival and reproduction—is obscured by the fact that "survival" is usually understood to mean the continuation of the life of a presently existing organism. If, however, we include X's being conceived as a contribution to X's survival (or perhaps the better term would be X's *life*), the equivalence becomes apparent. Dobzhansky notes: "Living beings must survive to reproduce, and must reproduce to survive in the following generation."[1]

The equivalence of the two ultimate goals is manifested in the common evolutionist aphorism that a hen is but an egg's way of producing another egg. The humor in this remark derives precisely from its shift from the ordinary perspective according to which an egg is a hen's way of producing another hen. Ignoring the rooster's role, the fact which both perspectives refer to can be stated neutrally: eggs are adapted to becoming hens, which are adapted to producing eggs, which are adapted to becoming hens, which are adapted to producing eggs, and so on. The difference, then between the reproductive-success perspective and the survival perspective is indicated by the following diagram:

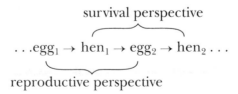

survival perspective

$$\ldots \text{egg}_1 \rightarrow \text{hen}_1 \rightarrow \text{egg}_2 \rightarrow \text{hen}_2 \ldots$$

reproductive perspective

What genetic selection favors is the chain of causation as a whole—i.e., the perpetuation of organisms of a certain genetic type. The difference concerns only whether this fact is to be summarized under the heading of "life" or "reproduction." But in reality, each goal contains the other: every living organism was produced through a successful act of reproduction, and every successful act of reproduction is performed by an organism that is alive and produces an organism that lives. To promote reproduction *is* to promote the formation of living descendants.

An example will illustrate the manner in which the facts often described in terms of the goal of reproductive success are translatable into the language and benefits to individual survival. Consider the action of reproduction itself. Viewed from the standpoint of reproduction as the ultimate goal, these actions are ends in themselves. How may they be seen as means toward the goal of individual survival?

The reproductive actions of the parents are understood as directed toward the goal of producing offspring. In the usual cases, however, this goal does not benefit the parents in any way. In fact, since the parents must expend considerable energy and take large risks in mating and reproducing, the consequences to the prospective parents' survival are negative: they would be vastly better off if they were able to refrain from reproductive behavior. On the other hand, the parents could not have come into existence at all, could not have lived, if *their* parents had not engaged in reproduction. The organisms of a given generation owe their lives to the reproductive behavior of their immediate parents and of all their ancestors back through the ages. It is the survival benefit (viz., being born) that the parents derived from past acts of reproduction in their ancestral line which explains their present acts of reproduction. Barring mutations that eliminate the gene complexes for reproduction, there is no way for a given organism to obtain the benefits of past acts of reproduction without having to repeat those acts upon their own maturity.

Hence reproduction is goal-directed by the individual-survival concept of genetic selection: reproduction is a self-generated action caused by the survival value *to the agent* of past instances of the goal. The goal here is the production of viable descendants; the past instances of the goal are the past instances of the production of viable descendants—including the production of the organism now engaging in reproductive actions. It is no objection that the present reproductive actions of the parents will not result in any consequent survival benefit to them: the proposed analysis is not phrased in terms of any *future* survival benefits which might result from a given action, but in terms of the action's *past* survival value. And it is true that the parents' present reproductive actions have been made possible by the survival value that past occurrences of that action have provided them. In this case, the past occurrences which benefited the parents were performed by their ancestors, but there is no reason why the survival value causing a given organism's action has to have been provided by *its own* past performance(s) of that action rather than by the past performances of that action by its ancestors.

By the same type of reasoning it can be shown that some apparently self-sacrificial behavior patterns in social animals actually serve the ultimate goal of individual survival. There is the well known example of the mother bird who lures a potential predator away from the nest by offering herself as "bait," thus diverting the predator from her offspring at the risk of her own life. This diversionary behavior has evidently been selected for in those species of bird which display it.[2] By the standard of reproductive success, the action is beneficial if the number of young which it allows to reach reproductive age is greater than the number of fertile mothers it eliminates. By the standard of individual survival, the action is beneficial if it allows individuals to survive to maturity which otherwise would have died as fledglings.

According to the individual survival perspective, the situation is understood as follows. To say that the diversionary behavior of mother birds is adaptive is to say it serves their lives. A mutation causing a female bird, when it reproduces, to protect its young in this manner has no survival value for this first individual. However, the action is beneficial in that the probability that its offspring will survive is rendered higher than the probability of survival of conspecific competitors. Over many generations, the increased probability of survival of offspring possessing the gene will cause that gene to spread throughout the population. Consequently, the reason why birds in subsequent generations perform the diversionary action lies in the survival value to each "altruistic" bird of past instances of that action, as performed by its ancestors.

Whenever it is the case that a given "altruistic" mother could not have matured (or could not have been born) but for the consequences of past instances of the diversionary behavior performed by *its* mother (or by any of the mothers in its ancestral line), then the present mother's performance of that action is caused by the survival value which it has derived from past instances of that action. Thus the diversionary action of the mother is not to be interpreted as an act of "self-sacrifice" (and certainly not as evidence that it "loves" its young more than its own life), but as a delayed "payment" for the survival value it has already received from the past instances of that action as performed by its ancestors. From the standpoint of individual survival as the ultimate goal, it is better to risk death at the age of parenthood than to have died as a fledgling (or never to have been born because one's ancestors died as fledglings)—and for sub-human organisms, whose behavior is genetically determined, this is the only alternative.

The topic of genetic selection is too extensive (and too specifically biological as opposed to philosophical) to be treated adequately here. But there are good grounds for the position that the ultimate basis of genetic selection may be viewed either as reproductive success or individual survival, the two being merely different perspectives on the same fact: differential perpetuation of alternative genetic units. Ontogenetic selection, on the other hand, is clearly based on individual survival as the ultimate goal. For a number of reasons, including simplicity and the fact that the concept of "reproduction" depends on the concept of "life," I prefer to view genetic selection from the standpoint of survival rather than reproduction, and this is the perspective I have adopted in this book. (I do not presume to suggest, however, that the survival viewpoint is preferable in biological inquiry—the choice of perspectives there is best judged by the biologists themselves.)

INDIVIDUAL VS. GROUP SURVIVAL

Next we must give some consideration to the question of whether the standard of genetic selection is the survival of the individual or the "survival" (i.e., the continued existence) of some group to which the organism belongs. I will defend the position that ordinary genetic selection operates by the standard of individual survival rather than group survival.

Genetic selection is the process by which individuals possessing certain genetic units are perpetuated over the course of many generations. The genetic units are possessed by the individual organism, not the group. A group of organisms (e.g., a deme, a species, or an ecosystem) does not possess genes, rather its members possess genes. The group has no existence over and above the existence of the individual organisms which compose it. The only way a group can "die" is by the death of all of its members; the only way a group can be maintained is by the maintenance of the lives of its members. But the members of most all groups (clonally reproducing insects aside) are genetically distinct: the content of their genetic programming is not coordinated by any mechanism possessed by the group as a whole (there are no such mechanisms[3]), rather it is determined by the individual success or failure of each organism in surviving and reproducing. Thus a gene allele which promotes individual survival at the expense of the group will not be countermanded by any "group entelechy" and may, in fact, be selectively favored to such an extent that it causes the extinction of the group.

Imagine, for example, a gene complex which caused one family in a densely populated species to become obligatory cannibals, eating their neighbors whom we will assume to be defenseless against such predation. Provided the mutant breed does not at first devour more of its own type than its

159

neighbors, such a breed could spread until there are only the cannibals left, at which point they themselves must become extinct (since they can survive only by eating each other). The group is defenseless against this sort of mutation: natural selection would not tend to eliminate it and preserve the group. The only hope for the group is that some of its non-cannibalistic members will evolve mechanisms of defense against the cannibals. But the group has no mechanisms for promoting the occurrence of the needed mutations.

This example is not entirely fanciful: it has been suggested that the extinction of the dinosaurs may have been caused by the emergence of a small, agile breed which lived on the eggs laid by other dinosaurs. In this case, there would be a high premium placed upon animals which could reproduce without laying eggs—i.e., upon mammals. This explanation of dinosaur extinction is speculative, but its conceivability argues against the view that genetic selection operates by the standard of group survival.

For all the foregoing reasons, the adaptations that *are* advantageous to the group are consequently better understood as advantageous to the individuals which compose the group. There are indeed cases in which an individual takes actions which, from a narrow perspective, seem to benefit the group at its own expense. However, Williams' book *Adaptation and Natural Selection* presents a sustained argument, going case by case, that these "altruistic" actions in fact are, in the long run, beneficial to the continued existence of individuals performing them.

(I consistently put "altruism" in quotes to indicate that the alternative of egoism vs. altruism, in its normal meaning, is inapplicable to animal behavior. "Human altruism presupposes conscious and free choice between alternative courses of action and is seemingly confined to a single species, man" [Dobzhansky, *Genetics of the Evolutionary Process*, p. 426].)

In principle, we can consider two general cases concerning "altruistic" behavior: either all the members of the group are genetically programmed to perform the behavior or only some members are. In the first case, the behavior may predictably result in the death of a certain percentage of the population, but if the behavior is beneficial to the group as a whole, the percentage of individuals dying because of the behavior must be less than the percentage of individuals which would die if they were not all programmed for its performance. This implies that the general probability of survival for each *individual* is increased when the behavior is universal in the group.

For instance, consider two hypothetical non-interacting groups: in group A the "altruistic" behavior is universal and in group B the behavior is not universal. By hypothesis, the behavior is beneficial and thus the probability of A's extinction is less than that of B. Over a number of generations, B will tend to become extinct while A continues to exist (assuming, of course, no change in the genetic composition of either group). This means that there comes a point at which the explanation of the existence of organisms in the A group lies in the consequences of past instances of that behavior—i.e., that each organism in the A group owes its very life to the fact that this behavior is universally performed. Hence, the action is not in fact "altruistic" since the gain to the individual of having that behavior universally built into the genotype of each member, including itself, outweighs the risk of death it faces when it is called upon to perform it.

An action cannot be considered "altruistic," in the sense of being beneficial to the group at the expense of the individual, if it is the case that were all individuals not programmed to perform it, the given individual could not have been born or could not have survived up to the time at which it must perform that action. In this case, the alternative faced by the individual is "some life vs. no life"—and it is strictly "egoistic" to "choose" the former.

In the second case, we assume that the individuals are not genetically identical so that some of the members of the group are programmed for the performance of the "altruistic" action while others are not. In this case, all members of the group benefit from the risky action which only some members perform. Here the "altruists" are at a competitive disadvantage and will be selected against if the action is truly "altruistic" (the cases of "kinship selection" cited by Haldane[4] are not truly "altruistic," since they involve the perpetuation of more genes similar to the agent's own than are lost in the performance of the action). If only some members of the group take the risks, but all members of the population benefit equally, then the "free riders" will benefit on the average just as much as the "altruists," but since the "altruists" have a greater mortality rate, they will tend to be eliminated from the group over time.

In either case we can see that in principle what selection preserves is the existence of individuals of a certain genetic type. If all the individuals in a group are of the same genetic type (and if there are no genetic changes), then anything which favors the continued existence over time of individuals of that type will necessarily favor the group which is composed of just those individuals. If the individuals in a group differ genetically, then what selection preserves is the existence of individuals of the genetic type with the greatest reproductive or survival fitness, regardless of its impact on the group. According to the principle of genetic selection, it does not "pay" to contribute to the continued existence of genetically different individuals. And any contribution to the continued existence of individuals genetically similar to the agent can always be interpreted as an "egoistic" rather than an "altruistic" act, since in this case the agent's own birth and/or survival was made possible by past instances of this kind of behavior.

The ultimate goal of genetic selection, then, is the survival of the individual of a certain genetic type, not the maintenance of the group.

As Williams states at the conclusion of *Adaptation and Natural Selection*:

> The species is therefore a key taxonomic and evolutionary concept but has no special significance for the study of adaptation. It is not an adapted unit and there are no mechanisms that function for the survival of the species.

> The only adaptations that clearly exist express themselves
> in genetically defined individuals and have only one ulti-
> mate goal, the maximum perpetuation of the genes
> responsible for the visible adaptive mechanisms.[5]

It appears that a conceptual confusion, rather than disagreement on the empirical facts, is largely (though not
entirely) responsible for the not uncommon belief that many
adaptations promote "the survival of the species" rather than
that of the individual. This confusion is obviated by the
previously discussed point that whatever favors reproductive
success favors the survival of the individuals successfully produced. Hence, any genetic unit contributing to reproductive
success is in the "self-interest" of each individual organism
possessing it (except the first organism in which it arose as a
result of mutation or sexual recombination). Again quoting
Williams:

> Somatic mechanisms, such as those relating to respira-
> tion, nutrition, etc., are usually said to be organic adapta-
> tions for individual survival, but reproductive processes
> are said to relate to the survival of the species. Certainly
> species survival is one *result* of reproduction. This fact,
> however, does not constitute evidence that species sur-
> vival is a function of reproduction. If reproduction is
> entirely explainable on the basis of adaptation for indi-
> vidual genetic survival, species survival would have to be
> considered merely an incidental effect [my emphasis].[6]

Biologist Ernst Mayr, after a review of the evidence
adduced in favor of group selection, reaches the same conclusion:

> No one denies that some local populations are more suc-
> cessful (and contribute more genes to the gene pool of the
> next generation) than others, but this is due to the aggre-
> gate success of the individuals of which these populations
> are composed, the population as such not being the unit
> of selection. . . .all attempts to establish group selection
> as a significant evolutionary process are unconvincing.[7]

BIOLOGICAL REPRODUCTION VS. MAN'S REPRODUCTION OF ARTIFACTS

It has been objected that my treatment of reproduction as an action having survival value for the organism produced opens the door to the conclusions that (1) a teleological explanation can be given for the existence of all man-made artifacts, and (2) that such artifacts as target-seeking torpedoes do exhibit goal-directed action after all. The first part of the objection maintains that, by the proposed analysis, a teleological explanation can be given for the existence and features of any man-made object, from Dixie-Cups to houses—a teleological explanation given in terms of the "value" of these features to the existence of the artifacts themselves (i.e., an explanation in terms of what Ayala calls "internal teleology"), not merely in terms of their intended use by the men who create these objects (i.e., not merely an explanation in terms of Ayala's "external teleology"). Let us consider the basis of this objection.

Take the case of the reproductive organs of animals. I have argued that such organs can be explained teleologically because the reproductive organs of the parents, while not aiding their own subsequent survival, have value-significance to their offspring, since the parents' possession of these organs is a necessary condition of their birth. Hence the offspring's possession of reproductive organs is explained teleologically by the contribution which their parent's possession of similar organs made to the offspring's life.

But in the same manner, it might seem that the specific features of a given Dixie-Cup (e.g., its shape, weight, tensile strength) can be explained teleologically. Such an explanation would be based on the fact that the value gained by the human manufacturers from past Dixie-Cups (i.e., the "ancestors" of the given cup) is what leads to the production of the given cup. In the case of living organisms, reproductive organs have value-significance because they lead to the production of similar organisms. Likewise, the features of Dixie-

Cups have value-significance because they lead to the production, by human beings, of similar "descendant" Dixie-Cups.

One reply that might be made to this objection is that the two cases are essentially different because, unlike living organisms, artifacts are not *self*-reproducing—their reproduction depends entirely on the actions of man. I am sympathetic to this approach, but I find it difficult to say just *why* self-reproduction is essential here. This reply also raises the counter-objection that some living organisms depend (although to a lesser degree) on the actions of other organisms for their reproduction (e.g., flowering plants depend on the actions of pollinating insects). Consequently, rather than pursue this line of approach further, I will turn to another and more basic way of answering the objection.

One important difference between the reproduction of artifacts and the reproduction of organisms is that only the latter has value-significance. Reproduction does not have any value-significance unless it leads to the production of an entity having its *own* goals (goals other than the production of still further "descendants"). Having been born is, retrospectively, *beneficial* to the organism produced, but it is beneficial only because that organism's life is, for independent reasons, a value to the organism itself. The organism's life is a value to it by virtue of the relation of its own independent, goal-directed actions to its survival. The self-sustaining actions of a living organism imply that its sustenance—i.e., its life—is a *goal*. In relation to this independently established goal, other factors, such as reproduction, can have value as means. But in the absence of any independent goal or end, one cannot identify something which facilitates a certain result as a *means*—that is, as a state having value-significance.

As argued in chapter 5, only life can be an end in itself because only life is sought for its own sake—only life is an action directed toward the perpetuation of itself. The Dixie-Cup's existence is not a goal for or value of the Dixie-Cup because it does not act to preserve its existence; the cup's existence, once it has been produced, is unconditional: it will remain in existence unless some external force acts to destroy it. The Dixie-Cup's continued existence has no value to it because its existence is not conditional upon its actions (it performs no actions).

For the same reasons, the motions of target-seeking torpedoes are not "for the sake of" the manufacture of future torpedoes. Because target-seeking torpedoes can further our goals, we manufacture new target-seeking torpedoes. But having been manufactured is not, even retrospectively, a value to the torpedo. The torpedo does nothing to maintain itself in existence—in fact, its motion leads to its destruction.

Reproductive actions can have value-significance only derivatively, in relation to the value-significance of the entity's self-sustaining actions. Even in the case of a living organism, the production of a fertilized ovum has no value-significance—at the instant of fertilization—to the zygote (nor, obviously, to the parents). It is only later, *retrospectively*, that we can say having been formed was a value to the offspring. And what allows us to say this is the fact that having been formed was a necessary condition for its subsequent self-sustaining actions.

It is only the production of an entity which values its own continued existence, i.e., an entity which acts to maintain its existence, that can, in retrospect, be said to have value-significance. Thus, the features and behavior of man-made artifacts do not exhibit teleological causation—except in terms of the ("external") goals of their human creators—despite the fact that their features and behavior lead men to manufacture similar new artifacts.

MALADAPTIVE PURPOSEFUL ACTIONS

Another general category of objections concerns the relationship between purposeful actions and goal-directed actions. I have argued that purposeful action is a sub-category of the wider class of goal-directed actions, purposeful action being distinguished by the causal involvement of consciousness. This implies that all purposeful behavior (at least for sub-human animals) brings some survival benefit which explains its existence. In many cases the survival value of purposeful action is virtually self-evident: eating is for the sake of nutrition, mating behavior is directed toward the goal of reproduction, pursuit and flight clearly benefit the survival of predator and prey, respectively. There are other cases, however, in which clearly purposeful behavior (at least for sub-human animals) has no immediate apparent survival value.

In principle, any purposeful behavior pattern that fails to provide a survival benefit represents a waste of the animal's limited energy resources and would tend to be eliminated by natural selection. For this reason, we are safe in assuming that any given behavior pattern normally found in a given species of animal is adaptive in its natural environment, even when this adaptive value is not immediately apparent. At least the burden of proof must rest with those who maintain that a given purposeful behavior pattern is *not* adaptive, since in that case a reason will have to be given why such behavior was able to resist the fine-toothed comb of natural selection. As Mayr has written, "Behavior is perhaps the strongest selection pressure operating in the animal kingdom."[8] Simpson observes:

> Human judgment is notoriously fallible and perhaps seldom more so than in facile decisions that a character has no adaptive significance because we do not know the use of it.[9]

In fact, the cases of purposeful animal behavior which *have* been carefully analyzed show a complete and intricate adaptedness that lends support to the generalization that all conscious behavior is adaptive. In an essay devoted to showing the need for increased study of the adaptedness of animal behavior, the pioneering ethologist Niko Tinbergen cites some examples of initially puzzling behavior whose survival value was discovered by detailed investigation:

> Lorenz (1931, 1935)[10] found that his captive jackdaws would repeatedly fly low over his head and wag their tails in front of him. This made him look at this behavior in wild jackdaws, and he discovered that it is a signal by which jackdaws stimulate members of the flock to join

them in flight. Another example: young ducklings and birds of several other species perform peculiar trampling movements when they are placed on a wet substrate (Tinbergen, 1962).[11] In the natural environment this is part of the feeding behavior: in shallow ponds the paddling stirs up motionless or concealed animals, which are then seen and eaten. . . .

A peacock butterfly (*Vanessa io*) has cryptically colored ventral wing surfaces. When at rest, these surfaces are exposed to view. When disturbed in cool weather, for instance by a human observer prodding it with a sharp twig, the peacock butterfly flaps its wings, thereby exposing the brightly colored "eyespots" on the dorsal surfaces and its fore- and hind-wings. While doing this the insect orients itself accurately in such a way that the surface of the wing is continuously turned toward the observer. Observations such as these led to experiments by Blest (1957)[12] that demonstrated that such eyespots scare off insectivorous song birds. . . .

E. Cullen's study of the kittiwake,[13] already mentioned, shows convincingly that such peculiarities as the building of a mud platform under the nest, the immobility of the chicks, the method of fighting by "bill-twisting," the absence of one of the threat postures generally found in gulls, the habit of guarding the nest even before eggs are laid, and numerous other characters are adaptive, and all corollaries of one antipredator device: nesting on sheer cliffs.[14]

In view of the general power of natural selection and of the extraordinary degree of adaptedness exhibited in the behavior patterns that have been analyzed, it would be conservative to estimate that of the total time and energy invested in conscious behavior, at least 99.9 percent involves behavior that is beneficial to the animal's survival.

Nevertheless, *some* purposeful animal behavior, even if constituting less than one-tenth of one percent of the total, is presumably maladaptive. If purposeful behavior is a sub-category of goal-directed action based on survival as the ultimate goal, how is one to account for the possibility of maladaptive conscious behavior?

One possible response would be to take the dilemma by its horns and flatly deny that maladaptive behavior can be purposeful, appearances to the contrary notwithstanding. This response, however, would evidence a desire to maintain a thesis, not to understand the distinctions found in nature. For instance, imagine that a group of horses were discovered that chased and killed whatever rodents came into view. Assume that after years of relentless study, zoologists could discover no adaptive value in such behavior, and in fact found that the rodent-chasing horses were generally less fit than entirely similar neighboring herds in which this behavior was absent. It would be perverse, under these circumstances, to refuse to describe the rodent-chasing behavior as purposeful, merely because it is not adaptive.

To take another example, the playful behavior of house cats in stalking and chasing their toys is a result of the cat's adaptation to the capture of the birds and rodents which serve as its food in its natural environment.[15] But even if no such explanation in terms of survival value were possible, we would hardly refuse to describe the cat's stalking behavior as purposeful. One would be tempted to say: "If this is not purposeful, what is?"

A proper understanding of maladaptive purposeful action is gained by recognizing it as a case of goal-directed action that fails to reach its goal. This presupposes that the behavior has a goal that is identifiable independently of the final state which the behavior happens to reach (in fact, as argued in chapter 5, this is a basic presupposition of any valid concept of "goal-directed action"). According to the proposed analysis, this kind of independent specification of an action's goal can be achieved: G is the goal of self-generated action A performed by organism O, if and only if O's present performance of A has as a necessary condition the survival benefit O derived from earlier instances of G as reached by earlier instances of A. The future instance of the goal may not exist and may not even be possible, yet the action by which the organism attempts to attain G is still goal-directed if it is causally explainable by the survival value of past G's. Thus, it makes sense, by the proposed analysis, to speak of goal-failure—i.e., to describe an action as directed toward a certain goal which it fails to attain in the present instance.

In general there are two pertinent reasons why a previously goal-attaining, adaptive action would fail to reach its goal in a given instance. One is a change in the conditions under which the action is performed—a change in the environment and/or a change in the organism.

For instance, consider the ontogenetic development of cryptic coloration in an individual British Peppered Moth (*Biston betularia*). Until air pollution blackened the trees in certain areas of Britain, this cryptic coloration had enormous survival value in protecting these moths from predation by song birds. As the trees became blackened, the continuing ontogenetic development of the old coloration was still explained by its *past* survival value, but due to a change in the environment, the development of the old coloration was no longer able to achieve its goal of camouflaging the moth.

In the case of the vermiform appendix in modern man, we have an example of the loss of adaptive significance caused not by a change in the external environment, but by a change in the behavior of the agent. The appendix is valuable in the digestion of raw meat which, presumably, was part of the diet of ancient man. With the shift to a diet of cooked meat, the ontogenetic development of the appendix has lost its survival value (and in fact figures as deleterious, due to the possibility of appendicitis). Nevertheless, the growth of the appendix in the embryonic development of contemporary individuals is still explained by its *past* survival value in maintaining the lives of remote ancestors of these individuals. Here the failure to attain the goal of raw-meat digestion is explained by a change in the dietary habits of man rather than a change in the environment.

The second general reason why a previously goal-attaining action would fail in a present instance concerns the triggering stimulus of the action. An action may be triggered by a stimulus which generally correlates with the existence of a need for the action, but this correlation is rarely absolute. In many cases the action may be triggered when no goal-attaining action is either needed or possible.

Consider, for example, the scratching behavior of dogs. This behavior has general survival value in that it tends to remove parasites (fleas, ticks, etc.) which are harmful to the dog. The behavior is triggered by the presence of an irritating object which causes an itching sensation. In some cases, however, the itching may be caused by something other than a harmful object, in which case the dog's scratching serves no survival need. For instance, dogs moving through woods pick up burrs on their coats. We may assume that these burrs pose no threat to the dog's health, but that they nonetheless produce the itching sensation and cause the dog to scratch at them. Under these circumstances, the goal of removing the harmful parasites is not attained because the action was triggered inappropriately. The dog's scratching behavior is certainly purposeful, and it is still goal-directed, even though in this case no survival benefit will result from the present instance of the action. The purpose of the dog's action could be described either as removing the burr or as relieving the discomfort of the itch; the biological goal of the action is something different, though related: removing harmful parasites.

This difference between purpose and goal in the case of animal behavior does not present any philosophical difficulty: it is only in the case of human action that the agent's recognition that an object has survival value can cause him to pursue it; animals, lacking the ability to understand abstract or long-range consequences of their actions (and their own mortality), are motivated by direct pleasures and pains without any conscious awareness of the ultimate biological function their behavior fulfills. For example, a hen's action of sitting on her eggs has the biological function of maintaining the proper temperature for the embryo chick's survival, but the hen is, of course, oblivious to this: her purpose is simply to cool herself, and the heat conducted away by the egg does just this.[16]

This analysis of goal-failure provides the means for explaining maladaptive purposeful action in animals. If the action is in fact purposeful, then it must have a psychological value-significance to the animal. We can summarize this psychological value-significance under the general headings of pleasure and pain. The manner in which the animal's nervous system is "wired" determines what kind of stimulus will cause pleasure or pain—i.e., what kinds of objects will serve as the triggering stimuli for the animal's conscious behavior.

And, as we have seen, the pleasure-pain mechanism itself has an adaptive significance which explains its existence: the pleasure-pain mechanism as such is teleologically explainable as a means to guiding the animal's conscious behavior toward states possessing survival value and away from states threatening the animal's survival. It is true that, in any given animal, the pleasure-pain mechanism may be "mis-wired," as it were, causing an animal to be triggered into purposeful behavior which has no survival value, or which is definitely harmful. Nevertheless, such purposeful behavior is still goal-directed in that the existence of the pleasure-pain mechanism *as such* is explainable by its survival value.

As in the case of the dog scratching at a harmless burr, an inappropriate triggering of a type of purposeful behavior does not imply that the behavior is not goal-directed toward the animal's survival. Even if an animal were neurophysiologically constituted in such a way that a given type of behavior that never under any circumstances benefited its life caused it to experience pleasure, this behavior could be viewed as a failure of the triggering conditions of the pleasure-pain mechanism.

The pleasure-pain mechanism as a general faculty exists because of the survival value it has provided animals in the past. Maladaptive purposeful actions, then, can always be treated as failures in the organization of the pleasure-pain mechanism, and consequently as instances of goal-failure, rather than as simply non-goal-directed.

To put it in the widest possible terms, although any given type of purposeful action may be maladaptive, the whole phenomenon of purposeful action in animals has a biological function which explains its existence. Purposeful action *per se* exists because it has a survival value in enabling organisms to adapt their actions more closely to a heterogeneous environment, and to change their behavior in an adaptive direction within their own lifespans (as opposed to vegetative actions which are genetically fixed and can be changed only by random mutation or chance gene recombination in reproduction).

(Note also that the elimination of pain, *per se*, has survival value. Pain has deleterious psychological and physiological consequences which, if not eliminated, impair the functioning of the animal. There is some evidence that long-term pain in animals causes physiological disturbances which can lead to health problems [e.g., ulcers]. Also, some studies have indicated that long-term frustration can cause something similar to neuroses in rats.[17] In addition, pain raises the general arousal level of the animal, and behavioral research has shown that over-arousal causes a disruption of behavior: there appears to be an optimal level of arousal beyond which the animal becomes increasingly inefficient in its actions.[18])

X

EPISTEMOLOGICAL ISSUES

PHILOSOPHIC VS. SCIENTIFIC ISSUES

In view of the role that the theory of natural selection plays in my account of teleology, one might wonder whether my thesis is overly dependent upon recent empirical findings of biological science.

Of course, there can be no objection to empirical content *per se* in philosophy: there is no other source of content, ultimately. Philosophy, like every other discipline, derives from and refers to observed facts. One would not question, for instance, my making use of the empirical facts that living organisms exist or that their actions can affect their survival.

Some empirical content is and must be contained in philosophical theories. On the other hand, philosophical theories should not be subject to the rise and fall of purely scientific hypotheses, since those hypotheses may come to be rejected in the light of new evidence. Moreover, since science builds upon basic philosophical principles (e.g., the basic axioms, the law of causality, the principles of logic), there is the danger of employing circular reasoning in using science to support philosophical conclusions. (An example of just this kind of circularity is the frequent attempt to use the principle of natural selection to refute skepticism. See pp. 191-192.)

Accordingly, I propose the following working criterion as regards improper empirical content in philosophical theory: a philosophical theory is open to the criticism of being overly tied to empirical science if it depends on highly derivative scientific knowledge, rather than facts available to common observation. As a rule of thumb, I propose we take as our standard of common observation the level of knowledge possessed by the Ancient Greeks. If a given philosophical position is not dependent on any facts unavailable to the Ancient Greeks, it is safe from the criticism of being tied to questionable or too derivative empirical premises. With this criterion in mind, let us look at the empirical assumptions of the proposed analysis of teleological causation.

I think it is clear that the principle I have called "ontogenetic selection" is not based on any advanced or questionable empirical premises. The basic premise of ontogenetic selection is that living organisms need to act in order to remain in existence, that their continued existence is conditional upon their successful performance of specific processes of self-generated action. This fact was certainly known to the Ancient Greeks. It did not take any advanced biological research to discover that plants, animals, and men need to do something in order to remain in existence, and that inanimate objects, such as stones, do not.

If this premise is accepted, ontogenetic selection and the teleology based upon it follow directly. The conditional nature of life implies that the present actions of living organisms have been made possible by the past instances of the goals these actions have attained. No scientific sophistication is required to see that, for instance, my present heartbeat can occur only if I am alive, and I could not be alive in the present if my heart had not beat in the past.

In fact, Aristotle records—only to reject—Empedocles' prefiguring of ontogenetic selection:

> Wherever then all the parts came about just what they would have been if they had come to be for an end, such things survived, being organized spontaneously in a fitting way; whereas those which grew otherwise perished and continue to perish, as Empedocles says his "man-faced ox-progeny" did.[1]

The principle of ontogenetic selection is, then, independent of the theory of biological evolution. Even if the species that now exist had always existed, as Aristotle believed, ontogenetic selection would still exist. It would remain true that the continued existence of each living organism, and hence its re-performance of its vital activities, is made possible by the organism's continued ability to obtain the items which satisfy its survival needs. Teleological causation in general and goal-directed action in particular are *necessary* consequences of the conditional nature of life.

The principle of *genetic* selection, in its Darwinian form, requires additional empirical premises, but none that were unavailable to the Greeks. Simpson accurately describes the reasoning on which Darwin's concept of natural selection is based:

> Darwinian natural selection was based on a few concepts all obviously true once they had been pointed out. After Darwin had pointed them out, honest biologists agreed that they had been extremely stupid not to see them before. (Just one naturalist, Wallace, did see them without help from—and without in turn helping—Darwin.) All organisms vary, some being more and others less fit for survival. Much of that variation is heritable by their offspring. All organisms tend to produce more offspring than can possibly survive in the long run. On an average, more offspring will survive from those parents whose heritable variations make them more fit. Therefore, on an average and in the long run, characteristics that adapt various lineages of organisms to the different environments available to them will accumulate progressively within them. Q.E.D. The conclusion follows from the objective facts of nature as inexorably as the proof of a theorem in Euclid follows from his subjective axioms.[2]

The following is Darwin's own statement of the principle of natural selection from *The Origin of Species*:

> Can it, then, be thought improbable, seeing that variations useful to man have undoubtedly occurred, that other variations useful in some way to each being in the great and complex battle of life, should occur in the course of many successive generations. If such do occur, can we doubt (remembering that many more individuals are born than can possibly survive) that individuals having any advantage, however slight, over others, would

have the best chance of surviving and of procreating their kind? On the other hand, we may feel sure that any variation in the least degree injurious would be rigidly destroyed. This preservation of favourable individual differences and variations, and the destruction of those which are injurious, I have called Natural Selection, or the Survival of the Fittest.[3]

None of the premises employed in the Darwinian argument are remote from simple observation (which, after all, was all that Darwin himself employed). Nor are any of these premises hypothetical—Simpson's description of them as "objective facts" is fully warranted.

Perhaps there is, however, an implicit premise which is not available to the naive observer, nor to the Ancient Greeks: that life has existed for a long enough period of time for genetic selection to have had appreciable effects. If the world had been created in 4004 B.C., or 5 minutes ago (per Bertrand Russell's skeptical suggestion), genetic selection would not account for the adaptedness of the species that now exist. Certainly Darwin's thinking was profoundly influenced by his prior familiarity with Lyell's *Principles of Geology* which defended the modern view of the antiquity of the earth and its living inhabitants.

Although the antiquity of life is an essential premise in the theory of *evolution* by natural selection, it is not required to defend the *teleological* significance of genetic selection in explaining the origin of adaptive features in the lives of individual organisms. For such a teleological explanation, all we need to know is that the present existence of an individual with inherited adaptive feature F was made possible by its parent's possession of F. Even if the world had been created in 4004 B.C., it would still be possible to give a teleological explanation, based on genetic selection, of, for instance, this robin's possession of wings. If this robin's parents had not possessed wings, they could not have survived and the present robin could not have been born. This robin was able to be born with the hereditary equipment for wing growth because its parents' possession of the same hereditary equipment enabled them to grow wings, and hence to fly, and hence to survive, and hence to reproduce, reproduction being in kind.

It is true that had the earth been created only recently, the remarkable precision in the details of the adaptive features of organisms could not be explained teleologically. Wings are, for robins, an outright survival requirement; the fine details of the arrangement of feathers on the robin's wings, however, are necessary to survival only in the context of competition over the very long run (see chapter 7 for a discussion of competition).

It is only when a feature would not have been present but for its adaptive value that a teleological explanation of its presence is justified. Consequently, features giving only a very slight competitive advantage may not explain the continued existence of organisms possessing them until after many generations have passed. Thus, the *number* of features whose origin could be explained by their survival value depends on the amount of time life has existed. But all those features without which an organism could not live to reproduce are teleologically explainable from the second generation on. To take the (absurd) limiting case, if the world had come into existence 5 minutes ago, none of the features of organisms could be explained by genetic selection—all would be caused by whatever or whoever created the world—but very many of these features could be explained by genetic selection as soon as the next generation appeared.

The neo-Darwinian, "synthetic" theory of evolution involves additions to, rather than any abandonment of, Darwin's theory. Since even Darwin's theory provides a concept of genetic selection strong enough for our purposes, it is not necessary to discuss the status of the empirical premises of the synthetic theory.

Empirical study is, of course, necessary to determine what are the goals of particular living actions, and here philosophy must defer to biology. And there are indeed cases of living action which are not goal-directed. Firstly, there are those living actions which are goal-directed with respect to certain of their results but not with respect to other concomitant results.

The heartbeat, under the description of "an action producing thumping sounds," is not goal-directed. Secondly, there are living actions which serve no goal whatsoever, though these are necessarily rare due to natural selection. In this category fall vegetative actions in their very first phylogenetic appearance as the product of a random mutation or chance genetic recombination. In addition, we may include such actions as the growth of tumors and the extension of the lower leg in the human patellar reflex.[4]

Likewise, some of the structural features of living organisms cannot be given a teleological explanation. Features lacking adaptive significance and features whose adaptive significance is not the cause of their presence in organisms are non-teleological. An example of the former kind might be the presence or absence of ear lobes in man. An example of the latter kind would be any feature in its first evolutionary appearance, before it has been exposed to either ontogenetic or genetic selection. For example, if we imagine that the genes for the opposable thumb first appeared full-blown in a single individual primate, then that particular thumb's ontogenetic development in that individual could not be explained teleologically.

It is also worth noting, in passing, that many teleological features do not have the specific function naively ascribed to them. Flowering plants, for instance, do not have their attractive colors in order to serve man's esthetic enjoyment. Their colors are generally explainable, however, as adaptations to the allurement of those insects which aid their cross-fertilization.

Thus, it is not "true by definition" that all living actions and features are teleological. Which living actions and features are teleological and which are not has to be discovered by empirical investigation.

DEFINITIONS AS CAPTURING FUNDAMENTALS

At this point it is appropriate to discuss some further epistemological issues involved in my general approach to the subject of teleological causation. I have assumed that for some types of concepts, we must refine our definitions as we gain more knowledge about the nature of the phenomenon being conceptualized. Our goal in this process is to make our concepts conform to the most fundamental similarities and differences found in the subject matter, and to have our definitions, consequently, phrased in terms of the most fundamental common characteristic(s). Specifically, I have been arguing that there are two distinct levels of causation in nature: the purely mechanical and the teleological (and, in particular, two levels of action: goal-directed and non-goal-directed)—and that the proposed analysis brings this fundamental distinction to light.

I do *not* hold that knowing my definition is a prerequisite for recognizing cases of teleological causation. It is quite clear, for instance, that people entirely ignorant of natural selection nonetheless speak of the heartbeat as being for the sake of blood circulation, and one can indeed maintain that a vegetative action has a certain effect as its goal while denying that the goal has any survival value. But I do not consider these facts to raise any objection to defining vegetative teleology in terms of natural selection and survival value.

In general, I do not accept the validity of Moore's "open-question" technique for challenging definitions (which is what the objection represents) for the same reasons which have been given against its use to attack definitions in ethics.[5] I am in general agreement with the criticisms of the open-question approach which have been raised by G. C. Field[6] and by W. K. Frankena.[7] One cannot refute a proposed definition simply by showing that one can identify instances of the class by their possession of characteristics other than those stated in the definition. Frankena observes that "if Mr. Moore's motto (or the definist fallacy) rules out any definitions, for example of 'good,' then it rules out all definitions of any term whatever."[8]

What I have attempted to do in deriving the proposed analysis of teleological causation is epistemologically equivalent to what the geometer does when he defines "circle" as "the locus of points in a plane equidistant from a given point" or the medical researcher when he defines "tuberculosis" in terms of the presence of a certain virus in the lungs. In each case, one starts with a pre-scientific awareness of a kind of phenomenon (goals, circles, or cases of tuberculosis) and later discovers a more fundamental property which accounts for the presence of the initially observed common characteristics. It is because every point on a circle is equidistant from a given point, the center, that the circle looks round; it is because of the presence of the tubercular virus in the lungs that the patient shows the superficial symptoms of coughing, weakness, etc. And it is because the end states of vegetative actions have been naturally selected for their survival value that one is drawn to describe those end states as the "goals" of those actions.

One can also observe that the progress toward a more advanced definition may result in the exclusion from the class of some phenomena which were initially included. For example, in terms of gross symptoms, emphysema of the lungs may resemble tuberculosis, but it is now classed as a separate disease; to the naive observer, the whale appears to be a fish, but biologists classify it as a mammal. Similarly, the actions of a target-seeking torpedo may, on a superficial level, appear goal-directed; the performance of a beneficial action in its very first instance is behaviorally identical to its performance in subsequent generations. In both cases, however, the proposed analysis directs us to regard the behavior as non-teleological, because it lacks the fundamental characteristics of teleological causation. This fact, by itself, should not dissuade us from accepting the proposed analysis as sound.

REDUCTION VS. ELIMINATION

It might be argued that the proposed analysis, in allowing us to define teleological concepts in non-teleological terms, merely provides the means for the elimination of teleological concepts from empirical biology. This position, however, presupposes the view that teleological concepts are epistemologically suspect—that if we *can* do without them, we should. While this attitude might seem to be based on considerations of parsimony—not wanting to multiply concepts beyond necessity—it is actually the expression of a residual anti-teleological bias, as will become evident in the following discussion.

This "eliminativist" view, the view that the proposed analysis should lead to the systematic elimination of teleological concepts, in effect begs the question of the validity of teleological concepts by posing the following dilemma: either teleological concepts cannot be defined, in which case they are unintelligible, or they can be defined, in which case they are to be eliminated in favor of their definitions. In either case the conclusion is the same: teleological concepts and teleological explanations have no place in empirical biology.

In detail, the dilemma is this: either teleological concepts can be defined or they cannot. If they can *not* be defined, then either we must regard them as nonsense terms, like "glyx," or else they have meaning which cannot be stated in a formal definition because, like "yellow," teleological concepts denote a primary fact; teleological concepts, however, are not *ostensively* definable in the manner of other basic concepts, like "yellow." One cannot *point* to an instance of teleological causation, at least not *qua* instance of teleological causation. Teleological concepts clearly pertain to abstract causal relationships not apparent on the strictly perceptual level (for instance, animals limited to the perceptual level can still be aware of the yellow color of an object, but could not be aware of the goal-directed nature of their actions nor of any of the survival functions served by their structural features). Thus, if teleological concepts cannot be defined, they are unintelligible and are to be ignored in empirical biology.

On the other hand, if teleological concepts can be defined, we again have two sub-cases: they will either be ultimately definable in non-teleological terms (i.e., by definiens free of teleological concepts) or definable in teleological terms only. If it is not possible to break out of the family of teleological concepts to define at least one of them in non-

teleological terms, then the whole family is collectively inde-
finable, and we revert to the preceding conclusion that teleological concepts are nonsense. It does no good, by way of
analogy, to define "glyx" in terms of "glip" if "glip" is only
definable in terms of "glyx" (assuming no ostensive definition
of either term is possible).

Alternatively, if as I maintain, teleological concepts *can* be
successfully defined in non-teleological terms, these concepts,
on the eliminativist approach, are to be replaced by their definitions. But then it is logically impossible for teleological concepts to have a role in biology.

More generally, if the eliminativist premise were correct,
then all concepts of any type whatever, except those denoting
basic sensations, should be eliminated. In other words, the
eliminativist premise is sensualism masquerading as empiri-
cism: if the fact that a concept can be defined, in a way which
is not ultimately circular, were grounds for doing away with
that concept, then only concepts standing for irreducible, primary sensations would remain.[9]

Clearly, the selective application of the eliminativist argument specifically to teleological, as opposed to mechanistic,
concepts is evidence of a lingering prejudice against teleology,
rather than representing the impartial application of a general epistemological principle. But if the arguments given in
this book defending the validity of teleological concepts are
sound, there is no reason for desiring to eliminate teleological
concepts from biological inquiry.

Consequently, the attempt to use the success of the proposed analysis to eliminate teleological terms, while retaining
mechanical terms, assumes teleology is more onerous than
mechanism—which is the very position the proposed analysis
attempts to refute. Presumably, all definable biological concepts could be eliminated in favor of their definitions, but
what would this accomplish, except the utter stultification of
biological thought?

Possibly the desire to replace teleological concepts with
their definitions results from a misunderstanding of the principle of reduction. The principle of reduction, understood in

proper terms, embodies the goal of the integration of scientific knowledge and the strengthening of derivative sciences by showing their principles to be consequences of more general theories contained in more basic sciences. It does not, however, imply the dissolution of the derivative sciences nor the elimination of the conceptual distinctions made in the derivative sciences. The reduction of chemistry to physics, for example, has not meant the demise of chemistry, nor of its specific concepts, such as "valence," "gas," and "acid." As Nagel states:

> The reduction of one science to a second—e.g., thermodynamics to statistical mechanics, or chemistry to contemporary physical theory—does not wipe out or transform into something insubstantial or "merely apparent" the distinctions and types of behavior which the secondary discipline recognizes.[10]

Why should it be otherwise in the case of teleological concepts in biology? The fact that goal-directed action is a special case of mechanical behavior does not imply that its distinction from other types of mechanical behavior is "merely apparent" and should be ignored. The distinction between (mechanical) processes caused by the survival value they have provided and (mechanical) processes caused by other factors is real and objective, even though it does not bespeak any "transcendent" principle operative in living organisms.

Just as the rejection of vitalism does not imply that there are no important differences between living and non-living entities, so a commitment to the position that all vegetative actions have physical causes, and that future ends are not causal agents, is compatible with making distinctions among processes according to their level of complexity.

Finally, it should be noted that a concept and its definition are not equivalent or interchangeable. Firstly, a concept and its definition are certainly not methodologically equivalent. For instance, the concept "rectangle" can function as a manageable unit of thought far more readily than can its definition, "a parallelogram with four right angles." And since the concept of "parallelogram" is in turn defined as "a quadrilateral with two pairs of parallel sides," we can see that the policy of always substituting definitions for the concepts they define would quickly lead to impossibly complex thoughts. The simple proposition "The diagonals of a rectangle intersect at right angles" would become: "The diagonals of a quadrilateral with two pairs of parallel sides having four right angles intersect at right angles." The original statement is clear; the expanded statement obtained by substitution of the definition for the concept it defines is obscure. (Imagine the confusion that would result by also replacing the terms "diagonal" and "right angle" with *their* definitions.)

Secondly, the concept conveys information beyond that which is explicitly stated in its definition. Specifically, the teleological concept of "goal" directs our attention to the similarities uniting naturally selected biological ends with the consciously selected ends of purposeful action. The whole thrust of this book has been to argue that this similarity is both objective and fundamental. This similarity of goals to purposes is *not* immediately apparent, however, if we substitute for "goal" its definition: an end state of a self-generated action whose earlier instances provided the agent with a survival benefit making possible the present action. In many cases the relationship of goals to conscious purposes may not be important, but in other cases it may well be, and the elimination of teleological concepts in favor of their definitions would at least make this relationship less evident.

Most fundamentally, the idea that a concept's meaning includes only the defining characteristic(s) of its referents stems from a theory of concepts that, though deeply entrenched in the history of philosophy, is wholly untenable. A concept, as Ayn Rand has argued, is a mental integration of the existents in reality to which it refers. The defining characteristic does not exhaust the nature of those existents (e.g., man is more than just rationality plus animality), and to limit the meaning of the concept to just those characteristics is an arbitrary procedure which results in cutting concepts off from reality. (For a thorough development of this issue, see Rand's *Introduction to Objectivist Epistemology*.)

DECIDING BETWEEN ALTERNATIVE
CONCEPTUAL HIERARCHIES

A crucial epistemological issue remaining is the objectivity of the conceptual hierarchy for which the teleologist argues. Suppose one grants that naturally selected vegetative actions have some attributes in common with consciously selected purposeful actions. There are nonetheless some important differences between them. Who is to say that the similarities outweigh the differences? In fact, does it even make any difference which "conceptual scheme" we adopt? Isn't it ultimately a matter of arbitrary preference how we classify processes, as long as the items we classify together have some similarity? If so, then we can view as the primary issue whether or not consciousness is involved, and regard naturally selected actions as merely analogous to consciously selected actions. What is wrong, in other words, with the following classification system?

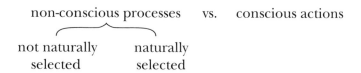

Note that the issue is taxonomic, not linguistic. The question does not concern the names we give to the categories, but how the categories are to be formed—i.e., which actions are to be joined as similar and which are to be separated.

I maintain as a principle of epistemology that classification systems are not arbitrary, and that the above classification is objectively wrong: the primary distinction must be drawn between living and non-living processes, not between conscious and non-conscious processes.

My grounds are that the similarities uniting vegetative actions with purposeful actions are *fundamental* while the differences between conscious and non-conscious actions are derivative. In appealing to fundamentality, I will be applying an aspect of Ayn Rand's theory of concept-formation:

> The process of determining an essential characteristic [employs] the rule of *fundamentality*. When a given group of existents has more than one characteristic distinguishing it from other existents, man must observe the relationships among these various characteristics and discover the one on which all the others (or the greatest number of others) depend, i.e., the fundamental characteristic without which the others would not be possible. This fundamental characteristic is the *essential* distinguishing characteristic of the existents involved, and the proper *defining* characteristic of the concept.
>
> Metaphysically, a fundamental characteristic is that distinguishing characteristic which makes the greatest number of others possible; epistemologically, it is the one that explains the greatest number of others.[11]

The role of fundamentality as a standard of classification is best explained by means of an extended example.

Suppose that we are attempting to determine a proper system of biological classification. We first consider the traditional Linnaean system of distinguishing the plant kingdom from the animal kingdom, then subdividing animals into phyla, subdividing the phyla into classes, and so on down to the species level, using the currently accepted criteria for each category. But we might also consider other bases of classification. We might take as our primary division, not plants and animals, but the two categories "stripes" and "non-stripes." A "stripe" would be any organism whose surface featured stripes, all others would be classed in the "non-stripe" kingdom. The "stripes" would include such organisms as are now called zebras, tigers, certain snakes, some plants, etc. It is true that all the organisms so classified would be similar—all the organisms of one kingdom would have stripes, all those of the other would lack stripes. And all the facts describable in our current system would seem at least to be describable in the new system as well. For instance, instead of saying "some plants are striped," as we do now, we could say "some stripes have the autotrophic form of metabolism" (plus whatever other properties characterize plants).

This new system of classification can be pushed to the point of total absurdity once we consider the possibility of defining subdivisions, such as the phyla, on the basis of equally superficial characteristics—such as whether the stripes are wide or narrow, black or white or chromatic, whether the organism's outer surface is hard or soft, etc.

Contrary to the claims of conceptual relativism, it is obvious that the Linnaean system is objectively superior to the classification of organisms in terms of being striped or not, having wide or narrow stripes, etc. The stripe/non-stripe system—assuming that anyone was actually capable of sticking to such categories rather than covertly reverting to the traditional classifications—would preclude the development of biological knowledge; in fact, it would turn biology into a hopeless muddle. Instead of zoologists and botanists specializing in plants and animals, we would have "stripists" and "non-stripists." A "stripist" would specialize in (what we now call) tigers, striped snakes, and striped plants, while ignoring (what we now call) lions, unstriped snakes, and unstriped plants. The cognitive advantages of specialization would be lost. If we add to the new disciplines, sub-fields based on such factors as the width and color of stripes, total scientific paralysis would result.

The reason for the propriety of the traditional system is, I submit, that its divisions are made in terms of *fundamental* similarities and differences, whereas the new system would be based on hopelessly superficial similarities and differences. The property of being striped is clearly too superficial to ground a primary division of living organisms—too little follows from the issue of whether or not a given organism is striped. Stripedness might serve as the distinguishing characteristic of a remote sub-category (e.g., of a breed within a species), but not as the primary division among all living organisms. In a given conceptual hierarchy or taxonomy, the primary division has to be formed along the most fundamental lines, with derivative issues being subordinated thereto.

> The requirements of cognition forbid the arbitrary grouping of existents, both in regard to isolation and to integration. They forbid the random coining of special concepts to designate any and every group of existents with any possible combination of characteristics. . . . In the process of determining conceptual classifications, neither the essential similarities nor the essential differences among existents may be ignored.[12]

It is true that "nature doesn't tell us" how to classify—in just the sense that "nature doesn't tell us" how to plow a field, build an automobile, or cure an infection. But the alternative to "revealed truth" is logically established truth, not subjective preference. Because "nature doesn't tell us," we have to use our minds to figure out enough about what is true of nature to achieve our purposes. In considering alternative taxonomies, the purpose is a cognitive one: the integration and advancement of knowledge. That purpose mandates certain standards of what is *right* and what is *wrong* in classification—e.g., the rule of fundamentality. If we want to understand nature, we must conceptualize it properly—we must divide nature at its joints. The joints, I maintain, are really "out there," even though it takes an effort of thought to locate them.

To classify actions properly, therefore, we have to ascertain the *fundamental* division among actions. Specifically, we have to choose between the criterion of "naturally selected vs. not naturally selected" and the criterion of "conscious vs. not conscious." To make this choice properly, we have to ask: which criterion is more basic? Are they co-equal or does one depend on the other? Although the answer in this case is far from obvious, I can indeed establish that the goal-causation resulting from natural selection is the fundamental.

My argument is that the faculty of consciousness is itself a biological adaptation—i.e., a feature whose presence in the higher organisms developed in evolution because of its survival value. As argued in chapters 8 and 9, the purposeful behavior afforded by conscious selection is not a primary fact, but rather represents a specific type of biological adaptation to the survival needs of more complex organisms. The whole phenomenon of consciously-directed action is thus dependent on the more fundamental phenomenon of natural selection, the phenomenon I am describing as non-conscious teleology.

If the form of goal-directedness resulting from natural selection is more fundamental than consciousness, our conceptual hierarchy should be organized to reflect this fact. Vegetative and conscious actions should be treated as two subdivisions of behavior shaped by natural selection, as in the following diagram.

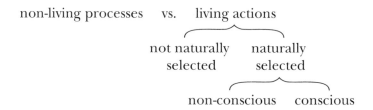

non-living processes   vs.   living actions

not naturally   naturally
selected   selected

non-conscious   conscious

To treat the conscious/non-conscious division as prior to the living/non-living distinction would be to deny the fact that adaptive non-conscious actions are more closely related to purposeful conscious actions than to the purely mechanical processes of inanimate matter—just as to treat "stripes" and "non-stripes" as the primary division of organisms would be to deny the fact that tigers are more closely related to lions than to striped plants.

(By categorizing conscious actions as a form of naturally selected action, I certainly do not mean to endorse that resurrected form of Social Darwinism known as "sociobiology." My point is not that human action is dictated by natural selection—a position I reject—but simply that consciousness as a faculty evolved for its efficacy in enabling organisms, including man, to modify their behavior in adaptive directions within their own lifespans, rather than being restricted to the glacial pace of evolutionary change. It manifestly does not follow from this that human beings, possessed of conceptual consciousness whose operation is volitional, are puppets of their genes. Nature endowed us with consciousness, including the conceptual level, because it makes possible enhanced survival; but how we choose to use our consciousness on the conceptual level is up to us. "Sociobiology," when applied to human action, stands convicted of the self-refutation that is inherent in all forms of psychological determinism: by the terms of the theory, the adoption of the theory is an act forced upon "sociobiologists" by their genes, rather than being an objective assessment of evidence. Human action is not "instinctual"; it depends upon one's thinking, deliberation, and decision—factors which cannot be claimed to be dictated by selection pressures, genes, or any other necessitating factor, on pain of invalidating the entire conceptual level and all of its conclusions. Any form of psychological determinism attacks the very possibility of rationality, while simultaneously claiming to be rationally arrived at.)

ATTEMPTS TO BASE EPISTEMOLOGY ON EVOLUTION

A related epistemological issue is whether or not one can appeal to evolution, not in regard to the theory of action, but in regard to epistemology itself. It has not infrequently been argued that the general validity of man's consciousness (on the sensory and/or conceptual levels) is assured by the fact that man's brain has been exposed to natural selection, and consequently that any types of brain not affording contact with reality have been eliminated. This argument is ultimately circular since the acceptance of the principle of natural selection already presupposes the general validity of one's mind. The principle of natural selection is a very derivative item of knowledge which rests on and requires the validity of a vast body of observations and on the validity of logical inference. Hence the attempt to justify observation and inference on the basis of natural selection begs the question on a grand scale.

(Furthermore, the argument evidences a misconception of the nature of selection itself, and is consequently fallacious even on its own terms. Selection does not guarantee that every member of an adapted species will inherit the adaptive genes. On the contrary, evolutionary theory is based on the observation that heredity is not 100 percent precise, that variations, most of which are maladaptive, occur. If so, then even if the *general* validity of man's mind could be somehow deduced from natural selection, how could the individual man propounding this argument ascertain that he, in particular, possesses the validity-conferring genes? If he does not, then his mind is not reliable, and all of his conclusions are incorrect, including the conclusion that selection operates to maintain the validity of man's mind.)

All of this goes to show not that man's mind is invalid—its validity is an axiom presupposed in any inquiry, thought, or discussion—but that any attempt to settle basic epistemological issues by appealing to conclusions from the special sciences grossly violates the necessary hierarchical order of knowledge.

XI

WIDER IMPLICATIONS

The fundamental similarity of vegetative and conscious action has important consequences for our approach to each. Each sheds light on the other: (1) the study of vegetative action is facilitated when, seeing its similarity to purposeful action, we approach it teleologically, and (2) our understanding of purposeful action is improved when, seeing its similarity to vegetative action, we approach it biologically.

IMPLICATIONS FOR VEGETATIVE ACTION

Identifying the goal-directedness of vegetative actions has two significant implications for biological research: it enables us to simplify the enormous complexity involved in such processes, and the concepts involved in the teleological approach—"survival value," "selection," "adaptation," etc.— make possible the grasping of abstract relationships and concrete implications that would otherwise be extremely difficult, if not impossible, to apprehend.

A. The economy of the teleological approach

Biological processes are generally the product of a large set of causal conditions. It is difficult to speak of *the* cause or *the* effect of, for instance, cellular respiration—there are many causal conditions and many effects. Adopting a teleological approach, however, means not considering all such causal relationships to be equally important. Teleological explanations of vegetative processes focus our attention upon those effects which further the organism's life. Is this selective focus valuable? Is there a sense in which an action's contributions to the organism's life is objectively more important than other of its effects? Yes: the contribution of the action to the organism's life figures not only as an effect, but also (in its past instances) as a cause of the action's occurrence.

Moreover, the survival contribution made by the action is not just *one* of the causal conditions of the action's continued performance, but it is normally the controlling condition, in the following sense. Where adjustment of conditions is possible, other conditions will normally be adjusted to the maintenance of the goal, rather than the goal being adjusted to the maintenance of any other condition. For instance, the contraction of an animal's leg muscles normally utilizes the oxygen carried in the blood to these muscles. Let us suppose that the relevant survival contribution made by these leg muscles lies in enabling the animal to run after the prey on which it feeds. But in prolonged muscular exertion, the animal is unable to supply oxygen to its leg muscles in a quantity sufficient to continue pursuit. If catching prey and the presence of oxygen were causal conditions of equal weight in leg muscle functioning, we would expect the animal to cease running when the oxygen supply became depleted. But, in fact, oxygen is a *means* to which running is the *end*—and we observe that in fact the animal's muscle continues to function by switching its basis of operation from oxidative breakdown of nutrients to anaerobic glycolysis.

> When exercise is raised to a strenuous level at which energy can no longer be provided in sufficient quantity through oxidation (because of the limit on the rate of delivery of oxygenated blood to the tissues), the muscles begin to supplement the energy supply by means of glycolysis.[1]

Faced with the alternative of the utilization of oxygen and the attainment of the goal, the muscle operation is adjusted to maintain actions toward the goal. Of course, the organism's ability to utilize alternative means to achieve a goal is limited by both its own nature and the environmental circumstances; the organism is not omnipotent. But whenever adjustment of behavior *is* possible, we observe that it is the end that determines the means, not the converse. The universality of this "plasticity" of means in relation to ends is noted in Davson's physiology text:

> Even the most primitive micro-organisms exhibit some control over their metabolic reactions, in the sense that, according to the physical and chemical conditions prevailing, the rate and often the nature of the reactions may vary.[2]

The variation in these metabolic reactions is, of course, not the simple mechanical kind in which different initial conditions lead to correspondingly different effects, but teleological plasticity in which different initial conditions call into play different mechanisms (or adjustments of the same mechanism) resulting in the maintenance of the *same* overall effect: goal-attainment.

This explains the reason for the simplifying value of the teleological approach to the study of vegetative processes. Since it is the goal of these processes that is the controlling causal factor in their occurrence, knowledge of the goal has a great explanatory power in regard to these processes. Firstly, we can explain the presence of some of the proximate efficient causes (e.g., aspects of the fuel and the mechanism) as products of past instances of the goal; secondly, we can explain other proximate efficient causes as being means to the achievement of the goal.

This is not to downgrade the value and importance of mechanical, and particularly biochemical, explanation in biology. According to the suggested approach, mechanical and teleological explanations do not compete. Any teleological explanation of a biological process assumes the existence of particular mechanical causes which operate to attain the goal in question; the teleological explanation merely does not *specify* what efficient causes are involved. Conversely, as argued in chapter 7, a mechanical explanation of a given biological process, if carried far enough, will state all the relationships which amount to a teleological explanation, even though, in a purely mechanical explanation, the terms in which teleological statements are made (e.g., "survival benefit," and "past instance of the goal") are not used.

Teleological and mechanical descriptions operate on different levels of analysis. To return to the analogy given in chapter 2, the term "World War II" bears the same relationship to the individual actions of the individual men involved in that conflict as the term "survival benefit" bears to the specific biochemical reactions involved in a given organism's survival. Just as the term "World War II" enables us to grasp easily complex relationships such as: "Germany was defeated in World War II," so teleological terms enable us to grasp easily complex relationships such as: "The zebra's coloration serves the goal of camouflaging the zebra to protect it from predators." In each case the statement implies the existence of, but does not specify, many precise details which add up to the overall relationship. Just as the science of military strategy would validly seek to analyze the details of the battles by which Germany was defeated in World War II, so the science of biology validly seeks to analyze the details of the means by which the zebra's protective coloration, in relation to the details of the structure and behavior of its predators, evolved. The two approaches are complementary: one gives a general overview which needs to be supplemented by specific details, the other gives specific details which need to be supplemented by a general overview.

The best general evidence of the cognitive value of the teleological approach in dealing with living actions is the fact that biologists, whatever their official theoretical commitments, employ teleological language constantly. Goudge observes that "both ordinary speech and technical treatises employ teleological language when describing what such internal organs as the heart, lungs, kidneys, etc., do."[3] Similarly, N. W. Pirie writes: "Biologists generally think teleologically but they then rephrase their ideas into a form that is both more acceptable to their fellow scientists and more productive of further work."[4]

W. T. Keeton in his textbook on biology, at first scores the teleological approach, calling it an idea "that can confuse scientific thinking" and stating, "teleological thinking has no place in a scientific study of animal behavior," but immediately admits:

> An attempt has been made in this book to avoid the most obviously and blatantly anthropomorphic or teleological expressions. But the book has been written in English, and English (like all human languages), having developed around human activities and human interpretations, inevitably reflects these.[5]

And we find that in fact his book contains many descriptions which are either explicitly or implicitly teleological—not merely in describing animal behavior, but even in describing "lower" processes. Consider, for example, the thoroughly teleological nature of the following passages.

> Our own tissues, particularly our muscles during violent activity, often need so much energy so fast that the oxygen supplied by breathing is insufficient. Under such circumstances, glycolysis provides the needed energy. Later, the oxygen debt is paid back by deep breathing or panting.[6]
>
> Taproots are frequently specialized as storage organs for the products of photosynthesis. Storage is a function of all roots, but particularly of taproots. Obviously, procurement of water and minerals and storage of high-energy organic compounds are not the only functions of roots; they also serve to anchor the plant to the substrate.[7]

It might be objected that the terms "need" and "function" are here being used neutrally, without implying teleology, in the above passages. There is, after all, a non-teleological sense of each term: "need" can mean simply "necessary condition" instead of "necessary for survival"; "function" can mean "usual effect" instead of "goal." But if this is so, why are only those "functions" and "needs" mentioned that relate to *survival?* In the non-teleological sense, the roots of plants have many other "functions"—e.g., they aerate the soil, they contribute to the eventual formation of petroleum deposits, they serve as food for sub-soil animals, etc. Likewise, the "needs" of the muscle, if "need" is interpreted as "necessary for some effect," are as limitless as the causal relationships into which muscles can enter. One could just as well speak of the "need" of the muscle to remain relaxed (or for the organism to die) in which case the "need" would be to eliminate energy, not to gain it, and the oxygen "debt" would become an "asset." Clearly, the implicit standard of "need" and "function" used in such writings is the survival of the organism—which involves just the kind of thinking that the teleological approach dictates.

As a final example from Keeton, I submit the following statement actually recommending the teleological approach (though not naming it as such) in regard to the study of plant leaves:

> As we briefly examine the structure of a representative leaf, keep in mind the following question: How do the structural features of the leaf make it an efficient organ for carrying on photosynthesis?[8]

As Dobzhansky observes, "Explanations in terms of adaptedness or teleology are not only appropriate but indeed necessary in biology, whereas they are meaningless in the non-living world."[9]

B. The theoretic power of the teleological approach

In addition, the specific concepts involved in the teleological approach make possible the grasping of abstract relationships, and laws, which would never be apparent from a purely reductivist, mechanical approach. The following are some examples of such biological laws (or hypotheses):

> 1. A given means of attaining a goal will be displaced by more efficient means, should they appear.[10]
> 2. "Altruistic" actions are favored only when the number of genes lost in the action (through death of the "altruists") is less than the number of similar genes saved (through the aided survival of the altruists' beneficiaries).

3. One goal of the coloration developed by animals subject to predation is to protect them against predation.
4. Reproductive isolating mechanisms are beneficial to potentially cross-breeding populations.
5. The pleasure-pain mechanism has the goal of motivating conscious behavior toward survival benefits.

Generalizations of this kind are the products of teleological thinking. Since all of the terms employed in them are *ultimately* definable by non-teleological equivalents, it is conceivable that each law could be translated into a non-teleological counterpart.[11] The possibility of such a translation is not the issue; the relevant point is that for identifying or applying the laws on this level, a mechanical approach is hopeless. Whatever may be true about the ultimate reducibility of biological laws, it is impossible to imagine a *biochemist* arriving at hypotheses such as those listed above from a study of the conditions and effects of biochemical reactions. The "translations" of these five and similar principles would simply be far too complex to be graspable or utilizable.

Each of these laws or hypotheses can be used to generate "testable predictions," as the positivists would put it—i.e., in application, the generalizations yield information about unobserved instances. Some examples follow.

1. If a given means of attaining a goal will be displaced whenever more efficient means appear, then there is reason to believe that the progressive loss of a given mechanism in evolution is a sign of a change in conditions, such that a more efficient means of attaining its end has appeared. In particular, the animals from which man has descended possessed specific mechanisms for the synthesis of ascorbic acid (vitamin C). In man and other primates, however, this mechanism is absent. Since ascorbic acid is necessary to prevent scurvy, we can assume that the loss of the mechanism for synthesizing it was made possible by the presence of adequate amounts of ascorbic acid in the diet of early man (and those other primates lacking the synthesizing mechanism). Thus, the elimination in man of the mechanism for synthesizing ascorbic acid when combined with the general hypothesis that less effi-

cient features tend to be eliminated (plus other non-controversial premises) has implications concerning the diet of early man.

2. The generalization that genetic selection acts to eliminate truly "altruistic" actions, but not those apparently "altruistic" actions which involve long-range gains to organisms genetically programmed for their performance, supports the inference that any newly discovered apparently "altruistic" action is actually beneficial to long-run survival. In fact, this hypothesis can be used to make very detailed inferences about behavior. For instance, consider types of birds that risk their lives by uttering warning cries which alert their neighbors to the presence of predators. The presence of this behavior, in the light of the general hypothesis, predicts what minimum percentage of the birds within hearing distance will on the average be close relatives of any individual bird. If, on the average, the families of birds of this type were widely dispersed, giving the warning cry would tend to be disadvantageous and hence eliminated by natural selection. From the presence of this behavior, the hypothesis predicts that a high percentage of the birds within hearing range will be closely related.[12]

3. The fact that coloration is generally adaptive for protection against predation could be used to infer, from its coloration, the kind of environment to which a given organism is naturally adapted, including the kind of predation to which it is normally subject. For instance, in the phenomenon known as Batesian mimicry, one species, e.g., a type of beetle, visually resembles an entirely unrelated species, e.g., a wasp.[13] In these cases the species that serves as the model (e.g., the wasp) possesses some form of anti-predator defense—a sting, a poison, or a repugnant taste—that predators learn to avoid; the imitating species (e.g., the beetle) lacks that defense, but predators often avoid it too, confusing it with the model species.

Knowing that the goal of the mimicry is to avoid predators in this manner allows us to make some interesting deductions. Suppose we observe that a given type of beetle closely resembles a certain type of wasp, and that this wasp uses its sting to defend itself against predation by robins (e.g., naive robins try to eat the wasp, get stung once, and avoid such wasps in the future). We then have good evidence to conclude that robins are also potential predators of that type of beetle, and even that the beetles are found in the same habitat as the wasps they mimic. Furthermore, the accuracy of the mimicking even gives us information about the relative population sizes of mimic and model in a given locale, for, as Mayr reports:

> If under special circumstances the model becomes rarer than the mimic, the latter may "go to pieces." As a result of relaxed selection the precision of the mimicry breaks down and a considerable part of the population consists of intermediates between the elsewhere sharply discontinuous types. This has happened in *Pseudoacraea* [a butterfly] on some islands in Lake Victoria. This phenomenon proves that the precision of the mimicry can be maintained only through the continued vigilance of natural selection.[14]

4. The fact that cross-breedings are not beneficial to either species involved can be combined with the fourth teleological hypothesis to predict the presence of reproductive isolating mechanisms in any two species where there is a significant potential for cross-breeding. Conversely, the absence of any specific isolating mechanisms in two species capable of cross-breeding is good evidence for the position that their natural habitats are geographically isolated.

In fact, the hypothesis affords some degree of expectation that if two genetically distinct animal populations previously separated geographically come into proximity, if cross-breeding is mechanically possible, reproductive isolating mechanisms will gradually come into existence through natural selection. Since selection is "opportunistic" rather than prospective, the emergence of such a mechanism is contingent upon the chance appearance of suitable genotypes. Nevertheless, experiments on the fruit fly have shown that under such conditions effective isolating mechanisms can evolve in just a few generations.[15]

5. The general hypothesis that the pleasure-pain mechanism has the goal of motivating conscious behavior toward survival benefits can be used to predict that any given object which is known to cause pain or produce pleasure is probably associated with biological harm or benefit, respectively. The following case cited in Dethier and Stellar, *Animal Behavior*, illustrates a situation in which taking this teleological principle into account could have saved a child's life:

> The case [is] of a three-year-old boy with an abnormal craving for salt. From early life, he always preferred salty foods and would lick the salt off bacon and crackers rather than eat them. When he was eighteen months old, he discovered the salt shaker and began eating salt by the spoonful. He learned very quickly to point to the cupboard and scream until he was given the salt shaker, and the very first word he learned was "salt." It turned out that his craving for salt had kept him alive, for when he was taken to the hospital for observation and placed on a standard hospital diet with limited salt, he died within seven days. At autopsy, it was learned that he had tumors of the adrenal glands and thus lacked the hormones necessary to reabsorb salt at the kidney. Only by constantly replacing salt lost in his urine did he maintain himself.[16]

IMPLICATIONS FOR MAN'S PURPOSEFUL ACTION

Some of the most interesting and philosophically important applications of the proposed thesis concern its implications for our view of man's purposeful actions, and more widely, our view of man's general place in nature. Heretofore I have explicitly refrained from discussing the complexities of distinctively human behavior, but it is possible and desirable to include an account of the teleological nature of forms of purposeful action that are unique to man.

Human purposeful behavior is frequently adaptive in the same manner as is the behavior of lower animals. The actions of getting an education, pursuing a career, finding a mate, etc., clearly are advantageous to one's survival. Even such actions as play and the pursuit of esthetic enjoyment can be seen as contributing to survival, insofar as they help maintain the individual's mental health.[17]

On the other hand, only man can act irrationally; only man can act in a consciously self-destructive manner. The lower animals cannot act irrationally because they cannot act rationally—not having the power to reason or think conceptually, the categories "rational" and "irrational" are inapplicable. Likewise, animals, unable to grasp the concepts of "life," "death," etc., cannot intend to harm their lives, to sacrifice their interests to some "higher" goal, or to commit suicide.

If purposeful action as a general phenomenon is to be viewed as an advanced form of goal-directed action, and if goal-directed action is based on individual survival as the ultimate goal, how are man's irrational and self-destructive actions to be understood?

The answer is provided by means of the same general approach that was taken in chapter 9 in discussing the problem of maladaptive purposeful actions performed by animals. If we take seriously the idea that man's consciousness is an adaptive faculty which evolved because of its survival value, we must regard the purposeful actions directed by man's consciousness as adaptive in general, even though they may have harmful consequences in any given case. Man has the ability to act purposefully because his survival requires such actions, even though he may act against his survival in any given case. To repeat the point made in chapter 9: to put it in the widest possible terms, although any given type of purposeful action may be maladaptive (i.e., irrational or motivated by a desire for self-sacrifice), the whole phenomenon of purposeful action in animals (including man) has a biological function which explains its existence. Purposeful action exists because it has survival value in enabling organisms to adapt their actions more closely to a heterogeneous environment, and to change their behavior in an adaptive direction within their own lifespan.

Despite its all-too-frequent use to achieve self-destructive ends, man's distinctive form of consciousness has proven its general survival value, by the fact that man has become the dominant and most eminently successful organism on earth. According to Dobzhansky:

> Judged by any reasonable criteria, man represents the highest, most progressive, and most successful product of organic evolution. The really strange thing is that so obvious an appraisal has been over and over again challenged by some biologists. . . .
>
> The evidence of the success of man as a biological species is ample and overwhelming. . . .
>
> The world population at the time of the Roman Empire is estimated to have been some 150 to 200 millions; around A.D. 1650 it was between 500 and 550 millions. The estimate for 1947 is about 2330 millions. The increase in number is, of course, not the only form of biological success. . . . However, man has become one of the few truly cosmopolitan species. He has penetrated into all parts of the earth's surface, and has established permanent habitation on all continents and major islands, except in Antarctica (and even there he manages to live for short periods of time). He has, accordingly, become exposed to every variety of geographic environment which the world has to offer, and he has adapted to these environments. But while animals and plants become adapted to their environments by modifying their bodies and their genes, man has remained the same and has to a considerable extent modified environments to suit his purposes and preferences, and has created completely new environments.[18]

In another work Dobzhansky attributes the biological success of man to the kind of purposeful action which his distinctive form of consciousness makes possible:

> The adaptive value of forethought or foresight is too evident to need demonstration. It has raised man to the status of the lord of creation.[19]

In *The Meaning of Evolution*, Simpson confirms this estimate:

> Man *is* the highest animal. The fact that he alone is capable of making such a judgment is in itself part of the evidence that this decision is correct. . . . In the basic diagnoses of *Homo sapiens*, the most important features are probably interrelated factors of intelligence, flexibility, individualization, and socialization.[20]

This same power of forethought, this same intelligence and flexibility that has accounted for man's success, necessarily carries with it the possibility of irrational, self-destructive purposes. Such purposes, however, represent the volitional misuse of man's tool of survival—that is, they are instances of failures in the functioning of an adaptive organ, rather than something sheerly non-teleological. Purposeful action has been selected for its survival value, even if an individual decides to use that capacity either irrationally or from a sense of "higher duty" to seek his own destruction.

The biological approach to human consciousness could have great significance for the science of psychology. Most of the attempts in this direction, however, have evidenced a quite non-empirical approach to man. There have been numerous attempts to treat man's psychology from an allegedly biological viewpoint while ignoring man's distinctive and fundamental means of survival: his ability to think, to conceptualize, to reason. Many authors (e.g., Desmond Morris[21]) appear to believe that man's rational faculty is biologically irrelevant and that what is biologically essential is his alleged innate aggressiveness, or his emotions, or his prolonged infancy, or his opposable thumb. In contrast to these approaches, it is well to note Simpson's statement:

> Human language is a virtually limitless, extremely complex, symbolic system capable of communicating anything sensed or experienced, including extreme abstractions and references to immediate or indefinitely remote past and future. Not even the rudiments of this kind of language occur in any other now living animal. It is an

absolute essential, probably the most important of all essentials, of the human condition.[22]

An authentically biological approach to human psychology would be premised on the recognition that consciousness in general and man's particular form of consciousness in particular—his rational faculty—have, like every other part of his makeup, a biological function. In short, man's mind evolved for its survival value.

Ayn Rand makes this issue vividly clear:

> Consciousness—for those living organisms which possess it—is the basic means of survival. For man, the basic means of survival is *reason*. Man cannot survive, as animals do, by the guidance of mere percepts. A sensation of hunger will tell him that he needs food (if he has learned to identify it as "hunger"), but it will not tell him how to obtain his food and it will not tell him what food is good for him or poisonous. He cannot provide for his simplest physical needs without a process of thought. He needs a process of thought to discover how to plant and grow his food or how to make weapons for hunting. His percepts might lead him to a cave, if one is available—but to build the simplest shelter, he needs a process of thought. No percepts and no "instincts" will tell him how to light a fire, how to weave cloth, how to forge tools, how to make a wheel, how to make an airplane, how to perform an appendectomy, how to produce an electric light bulb or an electronic tube or a cyclotron or a box of matches. Yet his life depends on such knowledge—and only a volitional act of his consciousness, a process of thought, can provide it.[23]

One common misuse of the biological perspective is the attempt to use the nature of evolution to prove what man's ultimate moral purpose is. I refer here to attempts to prove that man should subordinate himself to "goals" allegedly possessed by or revealed in the evolutionary process itself. (See, for example, *The Phenomenon of Man*, by Teilhard de Chardin, and Simpson's rebuttal in *This View of Life*.) It should be clear from the discussion of natural selection herein that evolution as such has no goals. Evolution has indeed produced organisms which have their own goals, but the process of evolution in itself is non-teleological. The evolution of fish was not "for the sake of" man, and man is not "for the sake of" anything beyond himself. To quote Ayala:

> The overall process of evolution cannot be said to be teleological in the sense of being directed towards the production of specified DNA codes of information, i.e., organisms. . . . Evolution can be explained without recourse to a Creator or planning agent external to the organisms themselves. There is no evidence either of any vital force or immanent energy directing the process towards production of specified kinds of organisms. The evidence of the fossil record is against any necessitating force, external or immanent, leading the process towards specified goals.[24]

And, as Simpson aptly puts it:

> Evolution has no purpose; man must supply this for himself. . . . It is futile to search for an absolute ethical criterion retroactively in what occurred before ethics themselves evolved. The best human ethical standard must be relative and particular to man.[25]

This does not imply to Simpson (nor to me) that an objective ethical system is impossible, but only that such a system is not to be based on some end transcending man's life itself. In this connection, I will close with these final words of Simpson's:

> Man has risen, not fallen. He can choose to develop his capacities as the highest animal and try to rise still farther, or he can choose otherwise. The choice is his responsibility, and his alone.[26]

APPENDIX

For an entire century, the significance of Darwin's achievement for the problem of teleology lay unnoticed. Then, within a period of about five years, three people independently came to realize that significance: Francisco Ayala, Larry Wright, and myself.

I came to the central idea of this work in November 1966, in the course of thinking about Ayn Rand's theory of value. One month later, I included my idea in an article on the Objectivist ethics that I wrote for a student publication:

> What distinguishes simple motion from action to gain a value? The seeking of values implies the existence of a causal connection between the action and the goal it is to achieve for the acting entity. In a goal-directed action, the process can be explained causally by referring to a goal to be achieved, and its relation to the acting organism. . . .
>
> Consider an ice cube and an earthworm both at rest on the top edge of a slightly tilted board. A sun lamp is focused on them. As the ice cube melts, it becomes more slippery until it begins to slide down the board and out of the lamp's rays. Likewise, the earthworm inches its way down the board away from the intense heat. Both the earthworm and the ice cube would have been destroyed had they not moved out of the heated area. The motion of each was caused by the effect of the dangerous heat on it. Haven't both acted goal-directedly? No. The direction of the ice cube's motion was not determined by the "goal" to be reached nor by its need to escape the heat, but by a random factor: the tilt of the board. Had the board been tilted another way, the ice cube would have moved closer to its destroyer; had the board been level, the ice cube would have melted passively. The earthworm, however, moved towards safety *because* it was safety.

> The earthworm is organized in such a way that certain stimuli cause it to approach, while others cause it to flee. What stimuli will produce what reaction is determined through the evolutionary mechanism of natural selection, by the nature of the stimuli in relation to the earthworm's life. Like all living organisms, it is built to approach that which generally favors its survival and avoid whatever harms it."[1]

I repeated that example and developed the point at greater length in a speech at City College of New York in April, 1968. And by the time I wrote my dissertation proposal in 1970, I had developed the full theory.

Ayala came to his quite similar theory from his work in biology. A geneticist, then working with Theodosius Dobzhansky, Ayala had been trained in Thomistic philosophy, which, in its secular side, follows Aristotle in stressing the final causation of living things. Ayala's 1970 paper, "Teleological Explanations in Evolutionary Biology,"[2] was called to my attention by Ernest Nagel while I was writing my dissertation, so I was able to include his views in the thesis.

LARRY WRIGHT

I unfortunately did not become acquainted with Larry Wright's work until after completing my dissertation; Wright's central ideas first appeared in two articles, published in 1972[3] and 1973.[4] The full presentation of Wright's position came in his book *Teleological Explanation*, published in 1976.[5] There, Wright reports that his theory "grew out of a critical reflection on a series of articles which appeared in the 1960s. These were papers (in chronological order) by Hempel, Canfield, Sorabji, Lehman, Gruner, and Beckner."[6]

Wright gives his analysis of goal-directed action by offering a "formula" for "S does B for the sake of G," where S is the "system" (e.g., an organism), "B" is the behavior, and "G" the goal. According to Wright's analysis, S does B for the sake of (attaining) G if and only if two conditions are met:

(i) B tends to bring about G

(ii) B occurs because (i.e., is brought about by the fact that) it tends to bring about G.[7]

In other words, to say an action is goal-directed is to say that it occurs because it tends to bring about the goal. His theory of goal-directedness, in other words, focuses on what I have called the issue of goal-causation—or, as he puts it, "consequence-etiology." He explains that his analysis

> offers consequence-etiologies as fundamental to teleological explanation. It is this which allows us to account for the forward-orientation of teleological accounts of behavior. When we say that teleological etiologies are consequence-etiologies, we are saying that the consequences of goal-directed behavior are involved in its own etiology: such behavior occurs *because* it has certain consequences. It occurs because it tends to achieve G.[8]

Unfortunately, Wright omits here what I regard as the crucial step in developing and defending consequence-etiology: the explanation of how it operates. How can it be that the future consequences play a causal role in the genesis of the present action? In his chapter on goals, Wright argues at length (and confusingly) that no such answer is necessary. His position is, in effect, that we can "just see"—by what he mischaracterizes as a "perceptual skill"—that certain behavior occurs because it tends to have certain consequences.

> The most important observation to be made about behavior directed toward a goal is that its goal-directedness is often obvious on its face. Many of our teleological judgments are as reliable and intersubjective as the run of normal perceptual judgments. . . . We can simply *tell* that behavior is directed, that it has a goal, sometimes even what the goal is.[9]

Not surprisingly, most of his examples of "obviously" teleological behavior involve the conscious behavior of higher animals—e.g., rabbits fleeing from hounds, predators pursuing prey, and birds building nests. But in these cases, there is no problem of backward-causation to be solved: the animal's present mental content causes its actions. Or more precisely, there is a problem only for philosophers so blinded by behaviorism that they refuse to acknowledge that rabbits see and hear the hounds that chase them.

What does Wright say about the case of vegetative action? Virtually nothing. No examples from the physiological level or concerning plants are given in his chapter on goals. The only example outside the realm of the higher animals is that of a flatworm, microstoma. He states that it is clear, on the basis of experimental evidence, that microstoma attack hydra "for the sake of" obtaining the hydra's nematocysts. But flatworms have a nervous system including a central ganglion that might well be considered a primitive brain; as I suggested in chapter 1, the flatworms are perhaps the simplest organisms to have attained consciousness (though nothing in Wright's discussion depends on their being conscious).

In his chapter on goals, Wright fails to come to grips with the main philosophical puzzle concerning teleology: how can the future consequences cause the present action? He seems to hold that because consequence-etiology is "intersubjectively testable," it is obviously real, and that is the end of the matter: no substantial philosophic problem remains. Contrary to his statement in the passage quoted above, Wright does not, in this chapter, actually "account for the forward-orientation" of teleological behavior, he merely insists on its existence.

In my own theory, consequence-etiology is explained by reference to the causal role of *past instances* of the goal. Wright, however, says, "the statement of the cause is always appropriately put in the future tense: that things were such that G will (tend to) ensue. And *this* statement concerns the state of affairs prior to B."[10]

In other words, Wright seems to be granting causal efficacy to the *present fact* that certain consequences *will* occur. But this is just a restatement of the problem. How can what *will* occur (but has not occurred) figure as a causal agent? Calling it a present fact that something will happen in the future does not bring that "fact" into the present. The "present fact that" phototropism will produce extra glucose is not something that can serve as a causal agent. In actuality, that extra glucose does not exist (to say it does not *yet* exist is a redundancy), and what does not exist cannot act. The "present fact" that something will result seems to amount to nothing more, in this context, than that human beings can predict its future occurrence, on the basis of their knowledge of what does exist now.

I must allow for the possibility that I am misreading Wright, since (a) he mentions the role of natural selection at the end of his chapter (but almost tangentially, in a discussion of reducibility), and (b) he appeals correctly to the role of past instances in his separate discussion of teleological "functions" in the following chapter. Furthermore, in treating of goal-directed action, his focus is almost entirely on conscious behavior and (inappropriately) on feedback mechanisms. He refers, in passing, to the goal-directedness of "lower animals and plants," which he describes, puzzlingly, as "very difficult to detect." But if he is concerned to offer an account only of conscious goal-directedness, I do not see that any discussion is needed beyond Braithwaite's quoted explanation: "the idea of the 'final cause' functions as the 'efficient cause'" (taking "idea" in the widest sense to include the various forms of mental content I discussed in chapter 3). But perhaps it was worth his pointing out that in purposeful action we are dealing with goal-causation ("consequence-etiology"), as preparation for his far more valuable discussion of teleological functions.

Wright agrees with most writers in the field in laying great stress on the difference between goals and functions (in the teleological sense of "function"):

> "Goal-directed" is a behavioral predicate. The *direction* is the direction of behavior. When we *do* speak of objects (homing missiles) or individuals (General MacArthur) as being goal-directed, we are speaking indirectly of their behavior. We would argue against the claim that they are goal-directed by appeal to their behavior (e.g., the missile, or the General, did not change course at the appro-

priate time, and so forth). Conversely, many things have *functions* (e.g., chairs and windpipes) which do not behave *at all*, much less goal-directedly. And behavior can have a function without being goal-directed (e.g., pacing the floor or screaming out in pain).[11]

There is, to be sure, a certain distinction between explaining the actions of organisms and explaining their structures, but to make much of this distinction is to do violence to the integration of structure and action that is of the essence of living organisms. In my theory, I explain the actions of organisms by their structures (mechanism plus fuel), which are then in turn explained by the results of past actions. My perspective has been: action-structure-action. Since the cycle repeats endlessly, one can equally well pick out the structure-action-structure phase and thereby give a teleological explanation of a structure. And this is just what we do when we explain the windpipe by stating its function.

As with actions, structures can be explained teleologically by two levels of selection: ontogenetic selection and genetic selection. On the ontogenetic level, the structure-action-structure account is: our windpipes exist with the structure they have because we have stayed alive (and hence have been able to maintain that structure); we could not have stayed alive if our windpipes had not possessed a structure that enabled us to breathe successfully. Graphically, we have the sequence: yesterday's windpipe structure → survival → maintenance of structure → today's structure.

Although it may not be obvious, the continued existence of the windpipe of a single, individual organism does need explanation. The windpipe is not an inert metal pipe; it is a collection of living cells. When the organism dies, so do those cells, and the windpipe begins to decompose. There is no comparable need for an explanation of a continued existence of, say, the gas line of an automobile. The gas line is not maintained by any process of self-repair, and leaving the car's engine turned off does not therefore mean the disintegration of the gas line. A car's gas line will normally be intact decades after the car has been abandoned. The case is quite otherwise for the windpipe of a corpse after only a matter of weeks.

Even bones, which last for eons, die with the death of the animal. A dead bone could not, for instance, function successfully if transplanted into the body of a living animal. On the other hand, the gross shape of the bone certainly remains, and it may well be that gross shape that we wish to have explained. And that shape, like many other facts about organisms, cannot be explained by the ontogenetic level of selection. But its explanation on the genetic level does exhibit the structure-action-structure pattern. The explanation of why, say, an animal's femur is shaped as it is would be given in terms of the geneotypic coding for the development of the required type of cells, in the required configuration, at the required bodily site. The explanation of why just *that* sort of coding is present, lies in the beneficial effects that having that specific kind of shape had for the animal's parents (and more remote ancestors). So the overall pattern is: parental femurs → parental survival and reproduction → offspring's development of femur.

We encountered the same chicken-and-egg optionality of perspective in discussing survival vs. reproduction as ultimate goals in chapter 9. Such cycles of reciprocating causation are what life is all about. Life *is* self-sustaining action—i.e., action that perpetuates itself. In the case of life, everything is simultaneously both means and end. The windpipe is a means to breathing, and breathing is a means to preserving the integrity of the animal, including its windpipe. Structure is for the sake of function, and function is for the sake of structure. I chose to focus on the action-structure-action sequence, but the whole truth is:

. . . structure → action → structure → action → . . .

I grant that it is linguistically odd to ask: "What is the goal of the windpipe?" But if we take it as: "What goal is the windpipe there to serve?" the proper answer refers to the same progression of cycling causes and effects as: "What is the function of the windpipe?" Getting air to the lungs is the function of the windpipe and getting air to the lungs is the goal the windpipe is there in order to serve.

I have argued that the ultimate source of all teleological concepts is our awareness of the purposefulness of our own actions. Note that the concept "purpose" does not make the structure-action distinction: "The (non-conscious) purpose of the windpipe is to deliver air to the lungs." The same linguistic generality is exhibited by other teleological expressions: "The windpipe exists for the sake of delivering air to the lungs." "The windpipe exists in order to deliver air to the lungs."

The fact that in English "goal" *connotes* the target of an action should not divert our attention from the fact that there is no essential difference between stating the cause of an action and the cause of a structure—not, that is, when we are talking about living action and structure. The linguistic differences between "goal" and "function" are simply irrelevant to the essential philosophic issues: what is teleological causation? how, in principle, does it operate? where is it manifested? And in biological investigation, whether we are seeking the teleological cause of an action in terms of the goal it attains or the teleological cause of a structure in terms of the action it facilitates, the issue is: which effects of the action or structure have caused it to be selected over its alternates?

Further evidence that the goal-function distinction is only one of perspective, without philosophic or biological significance, is furnished by Wright's own analysis of "The function of X is Z":

  (i) Z is a consequence (result) of X's being there,
  (ii) X is there because it does (results in) Z.[12]

This is the same analysis Wright gave for "goal." The equivalence of the two analyses becomes more obvious when we put the earlier analysis into a grammatical form paralleling that of the "function" analysis, and use the same lettering system:

The goal of X is Z if and only if:
  (i) Z is a consequence (result) of X's being performed,
  (ii) X is performed because it does (results in) Z.

We can make one more simplification. In each case, part (i) of the analysis seems redundant; it is entirely covered by part (ii). Eliminating them, and the parentheticals, we are left with the following comparison:

  Functions: X is there because it results in Z
  Goals: X is performed because it results in Z.

Thus, Wright's own analyses of "goal" and "function" turn out to be identical, the difference being only what sort of thing the variables stand for: in the analysis of functions, "X" denotes a structure and "Z" an action; in the analysis of goals, "X" denotes an action and "Z" an outcome of the action (which, in turn, promotes the organism's survival).

But Wright's further discussion of functions is far superior to his discussion of goals. For in regard to functions, he solves the major problem: how consequence-etiology is possible without the implication that the future causes the present.

The point of this chapter is to say something helpful about function attributions and functional explanations in instances not underwritten by human design or intent. How are we to understand such claims as, "the heart beats in order to circulate the blood," or "the function of a porcupine's coat of quills is to protect the little beast from predators"? Given a background of natural selection, these cases—natural functions—can be understood in the very same terms as conscious functions, namely in terms of (F) [his analysis of "function," above], with only the slightest change in nuance. For just as conscious functions provide a consequence-etiology by virtue of conscious selection, natural functions provide the very same sort of etiology as a result of natural selection.[13]

The teleological function of an item, Wright observes, is not merely what something happens to be good for. The porcupine's coat is "good for all sorts of things that are not 'What it's there *for*'." Rather, "the function of the coat is that particular thing it's good for which explains why it's there."[14] And he lays great stress, as I do, on the pivotal epistemological role of "the difference between X's function, on the one hand, and something X does only 'by accident,' on the other."[15] This leads him to hold, as I do, that:

> first mutations are accidents in the sense relevant here. The first occurrence of a useful physiological structure cannot, on the evolutionary account, be attributed a function: it does not have a consequence-etiology, it did not get there because of what it does. Such structures can be given natural functions only after some consequences of its being where it is becomes part of its etiology and explains how it came to be there.[16]

Wright goes on to discuss the time-reversal issue in more detail. "Why not put the etiological contention in the past tense and avoid the problem altogether?" he asks. "Why not say X is there because in the past X's have done Z?"—which is similar to my way of putting it.

> The reason is twofold and very important. First, use of the past tense in this way blurs the distinction between functional and vestigial organs, which is worth some pains to avoid in this context. Both kidneys and appendixes are there because of the function they had in the past; only kidneys are there because they do what they do, which is to say only kidneys (still) have a function. In general, when we explain something by appeal to a causal principle, the tense of the operative verb is determined by whether or not the principle itself holds at the time the explanation is given. Whether the causally relevant events are current or past is irrelevant. To put it any *other* way would be misleading.[17]

In other words, making explicit reference to past instances, as I do, would carry a misleading implication: it would imply, Wright holds, that the causal relationship is over and done with. Suppose we say: the kidney's goal or function is to filter metabolites from the bloodstream. Wright maintains that we cannot treat that as a shorthand way of saying: the kidney is there because, in the past, kidneys have filtered metabolites from the bloodstream. Why not? Because it sounds like we are saying, absurdly, that kidneys no longer filter the bloodstream. The contrast with the appendix seems to support this: the appendix is there because in the past appendixes (say) aided in man's digestion of raw meat (but they no longer do).

But, as stated, this line of argument sacrifices philosophic accuracy to linguistic form. The *truth* of the matter is that it *is* the past instances of bloodstream purification that caused, through their contribution to the organism's survival, the present kidney and its present activity. And it is the past instances of digesting raw meat that caused the genes for appendix growth to be naturally selected. The future value of the kidney simply cannot figure into the explanation of its present existence. The functional explanation here is an evolutionary explanation, and evolution cannot anticipate the future.

Suppose we grant that the grammar of "X's are there because in the past X's did Z" could be misread in a given case. Suppose it is indeed open to the implication that X's no longer do Z. That unwanted implication has nothing at all to do with either the philosophic or the biological issue at hand. The linguistic point concerns whether or not we read the statement to imply that X's no longer do Z. But the philosophic issue is whether X's are now present because of past effects of past causes (which they are) or whether they are now present because of the future effects of present causes (which they are not). Both kidneys and appendixes are, in fact, susceptible of teleological explanation: appendixes, like kidneys, are now present because of what they have done for survival—it is only that what appendixes do is no longer valuable.

There is nothing scandalous in saying "X's are there because in the past X's did Z"—provided we are prepared to back up that language with the more precise account: organ-

isms having X's were able to do Z in the past; doing Z benefited the organism; that benefit has made possible the continued existence of organisms with X's.

Wright's worry about the linguistic implications of using
the past tense can be taken care of by purely linguistic
means. We can say: "kidneys are there because kidneys have
purified the blood—and, P.S., they continue to do that." (For
appendixes, we would simply negate the corresponding P.S.)
We could cast the statement in the future tense: "kidneys will
continue to exist because they are now purifying the blood."
We could make a generalized statement: "at any given time,
the kidney's purification of the blood is necessary to the existence at a later time of organisms with kidneys." The possibility of such rephrasings should be sufficient to separate the
linguistic and factual issues. Consequently, we can make
explicit reference to the causal role played by past instances
of a goal or function, and indeed we must make such reference in order to solve the problem of "future causation."

The main differences between Wright's theory of teleology and my own are epistemological. Wright's book is Kantian
in several ways: in his view of objectivity, in his pragmatism,
and in his acceptance of the analytic/synthetic dichotomy.

One of Kant's most insidious legacies is the socialization
of objectivity—the substitution of the collectively subjective
for the objective. The base of this switch, for Kant, is the
notion that true (noumenal) reality—"things as they are in
themselves"—lies forever hidden beyond the veil of appearances (phenomena). Although man's knowledge cannot ever
succeed in grasping the real, external object, so that actually
objective cognition is impossible, we may avoid total cognitive
collapse, Kant assumes, by adopting as our cognitive standard
those distortions that are inherent in the mind's "synthesizing
activity," as opposed to merely idiosyncratically personal distortions. Since genuinely objective knowledge has been ruled
out of Kant's court, he is left with the choice between private
and shared delusions. This choice itself can only be subjective
(though Kant ignored this); Kant opted for the latter: the subjective choice of shared subjectivism now being styled "objec-

tive."

Wright almost certainly does not accept the baroque and contradictory Kantian groundwork of this redefinition of "objectivity," but throughout his book, "intersubjective agreement" on the application of teleological terms is the ultimate standard to which he appeals.

Pragmatism is a Kantian offshoot, and Wright's pragmatism shows up in his tendency to bypass concern with explanation, analysis, and validation, in favor of "it works." In defending his account of goals (B occurs because B leads to G), Wright eschews the need for causal explanation in favor of "testability." He conceives of a kind of Mill's-Methods testing of constant conjunction, rather than an identification of what in the nature of the entities causes their action, as all that we need to validate teleology:

> Now to demonstrate to a skeptic that the particular B that occurs, occurs *because* it is the one that will bring about G is like showing, say, that it really was the removal of the coil wire which caused the car's motor to stop. . . . In general what is required is the elimination of alternative accounts of the phenomenon. . . . If we suspect that there is something about the way in which the coil wire is removed, which is independently killing the motor, we may try removing the wire in different ways or using different devices, and we could move the ignition circuitry around to change as many incidental relationships as possible.[18]

To illustrate further his view of objectivity, he applies it to a mathematical process:

> It does not matter in the least what bizarre things go through your head when you are taking square roots, so long as you always get the answer right you have an objective skill.[19]

But, in fact, an *objective* skill is precisely what you do *not* have, until you can explain why the method works—i.e., until you can logically connect the process you go through to those facts of reality which cause it to work. You may have a skill all right, but the precise need for the concept "objective" here is to differentiate rules of thumb, rote memorization of statements by "authorities," lucky guesses, folk recipes that work only in a limited domain, and downright fraud, from that which is guided by an awareness of the nature of the object in question. (And there is a certain question-begging here in the assumption that "bizarre" methods *could* "always" work.) If something "works," the immediate question for the man of science, as opposed to the pragmatist, is: *why* does it work? As Aristotle put it:

> Men of experience know that the thing is so, but do not know why, while the others know the "why" and the cause. . .thus we view them as being wiser not in virtue of being able to act, but of having the theory for themselves and knowing the causes.[20]

A method that merely seems to work is one that has not yet been validated as *objective.*

The Kantian-pragmatist view of objectivity as intersubjective testability underlies Wright's "dead-metaphor" approach to the whole issue of teleology:

> It will be the central contention of this essay that teleological expressions in most nonhuman applications represent dead anthropomorphic metaphors. The cases in which it is easiest to show the objective (intersubjective) application of these expressions are cases of elaborate goal-directedness in animals and mechanisms. But there are other sorts of cases that qualify.[21]

A "dead" metaphor is one which at an earlier stage in its linguistic history actually evoked a comparison—actually functioned as a metaphor—but whose constant use has made it so stale that it now functions as a literal designation. For instance, when scientists today use the term gravitational "attraction," we are no longer aware of the metaphor—i.e., of the fact that this term originally made a comparison to romantic attraction.

The metaphor has long since died. Forces of "attraction" and "repulsion" now have literal meanings ("toward" and "away" from) despite their origins as anthropomorphic metaphors.

Thus far, Wright's thesis that teleological concepts represent dead anthropomorphic metaphors seems fairly equivalent to my point that although human purposeful action is the original and paradigm case of teleological concepts, those concepts may now validly be expanded to cover a wider scope. But my point is epistemological, and depends on an entire (Objectivist) view of the basis of concepts and the conditions of their proper expansion in the light of wider knowledge (see Ayn Rand, *Introduction to Objectivist Epistemology*, chapter 7). Wright's view is more linguistic and proceeds from his Kantian-pragmatist view of objectivity. Accordingly, Wright holds that moving beyond metaphor to an expanded literal usage requires no epistemological validation. In particular, it does not require that the dead metaphor "must be capable of literal paraphrase or translation"[22] of the kind I gave above for forces of attraction and repulsion. He admits that:

> There is something compelling in the position that if a metaphor has empirical content, if it expresses a substantive *proposition*, then that content, that proposition, must be expressible in *non*metaphorical terminology.[23]

But, short-circuited by the idea of intersubjective testability, he immediately proceeds to deny the need for such literal translatability—i.e., for any validating definition of the concept in its expanded sense:

> If "explosive temper" and "green recruit" are as intersubjective and unproblematic in use as [such nonmetaphorical terms as] "magenta," "cloudy," and "loud," that should be enough to show parallel descriptive force. . . .
>
> In death a metaphor becomes its own literal paraphrase: it takes on objective criteria of application just like any other literal expression. . . . So the mortality of a metaphor is the index of its substantive content: if it can pass the objectivity tests, the results of those tests can become application criteria for the term.[24]

In other words, we do not need a definition of "goal" in order to objectively validate calling nonhuman actions "goal-directed," because everyone will agree on which actions we will extend that originally anthropomorphic term to cover. (Wright acknowledges that agreement on the extended use will require that the people involved be aware of the empirical data, since there are scientific facts involved—e.g., what the teleological function of a given organ is.)

This view appears to be an application of the Wittgensteinian injunction to look for a term's use rather than its meaning. If so, I would oppose it root and branch. Beyond the level of concepts of perceptual concretes, in the realm of abstractions derived from earlier abstractions, definitions are absolutely required. Definitions state the *essential* characteristics of the concept's referents—i.e., the *fundamental* characteristic which causes and explains the greatest number of the referents' other characteristics (see *Introduction to Objectivist Epistemology*, chapter 5). Mere "application criteria" all too often represent classifications made on the basis of superficial similarities among phenomena which are fundamentally different and which, accordingly, should not be lumped together. Such conceptualization by non-essentials is invalid, as I argued in detail in chapter 10, using the example of classifying organisms into "stripes" and "non-stripes." The fact that many people can (or think they can) apply a term without having a definition does nothing to establish either that term's objective validity or its referents' essential nature.

But the long excursion into the theory of concepts that would be necessary fully to distinguish definitions from criteria of application proves unnecessary in regard to the case at hand: the proper scope of teleological terms, such as "goal." For it is clear that there is no "intersubjective agreement" to be had: Wright wants to extend teleological concepts to cover man-made mechanisms, such as "target-seeking" torpedoes. I regard that extension as unwarranted—as a metaphorical usage of "seeking." Wright disagrees. The issue is not to be settled by poll-taking in regard to linguistic usage: in the body of this work, I give extensive reasons to defend my claim that no non-living process is teleological. The "intersubjective *disagreement*" goes further. At the conclusion of his chapter on goals, Wright maintains that "the behavior of electrons in what is called 'Pauli interactions' is unquestionably[!] teleological."[25]

The inclusion of man-made mechanisms and "other sorts of cases" bespeaks the crippling effect on Wright's approach of the third Kantian legacy: the analytic/synthetic dichotomy. (Although Kant, of course, did not originate this dichotomy, he institutionalized it.)

In commenting on Ayala's theory of teleology, Wright states:

> He defines "utility" in living organisms by "reference to survival or reproduction. A structure or process of an organism is teleological if it contributes to the reproductive efficiency of the organism itself, and if such contribution accounts for the existence of the structure or process" (p. 13 [of Ayala's "Teleological Explanations"]). And this seems to suggest that it is impossible by the very nature of the concepts—logically impossible—that organismic structures and processes get their functions by the conscious intervention (design) of a Divine Creator. This, I think, is an analytical arrogance. I am, personally, certain that the evolutionary account is the correct one. But I do not think this can be determined by conceptual analysis: it is not a matter of logic. . . . The consequence-etiological analysis [as advanced by Wright] begs no theological questions: the organs of organisms logically could get their functions through God's conscious design.[26]

Unfortunately, the charges of "analytical arrogance" and question-begging must actually be laid at Wright's door. For what Wright is implicitly maintaining is that the question of God's existence is outside the province of logic! Wright's acceptance of the analytic/synthetic dichotomy's corollary— the logically possible/factually possible dichotomy—has led him to sweep aside the centuries of philosophic arguments purporting to demonstrate the *illogic* of the belief in an omnipotent, omniscient, nonmaterial Spirit. The point is not that the atheist case is correct (though it is); the point is that Wright is here claiming that (a) the theist-atheist dispute is in principle beyond logic (which is analytical arrogance), and (b) that we can waive dispute on the metaphysical fundamentals involved in the question of God's existence and proceed to pursue highly derivative, technical issues, such as the issue of teleology (which is question-begging on a grand scale).

Objectivists maintain that the idea of a Divine Creator contradicts every philosophic axiom, and thus undercuts all the rest of philosophy—and all the essentials of man's knowledge. The self-sufficiency of existence, the absolutism of causality, the essential dependency of consciousness on external reality, the supremacy of reason—these are just the "highlights" of what the notion of God wipes out. Thus the issue of atheism vs. theism—which means the issue of reality vs. the supernatural and reason vs. faith—cannot be "bracketed" while going on to discuss subordinate issues, such as teleology.

Ironically, Wright's complacency toward theistic accounts of teleology, which he sees as necessary to avoid begging questions, begs the question directly at issue between us. For according to my theory, the organs of living organisms logically could *not* get their functions through God's conscious design. God is supposed to be immortal, indestructible, impassive. Therefore, if the alternative of life or death is the basis of teleology, God is logically debarred from having any purposes. Since he cannot be affected by anything, since he has nothing to gain or lose from any state of affairs, he cannot be a Designer. To put it simply, how can God care what happens in the world, since what happens cannot affect him in any manner? Not being able to care, he can seek nothing and design (in the teleological sense) nothing.

Of course, to defend this point adequately, I would have to show why God's "stake" in the world could not be a purely emotional one, a stake not derived from any existential alternative. And that would take me back to prior issues concerning the dependence of emotions, as states of consciousness, upon phenomena of the external world. (The fundamental issue is what Ayn Rand termed the primacy of existence vs. the primacy of consciousness.) My concern here is to show that keeping an "open mind" about fundamentals is actually a form of begging all the basic questions.

The pragmatist standard of "intersubjective testability" fosters the uncritical acceptance of conventional assumptions, such as that it "of course" makes sense to talk about God's having purposes. But if my analysis is correct, then the extension of the human concepts "purpose" and "design" to an immortal being's activities, despite its near-universal acceptance, is cognitively meaningless. The conventional assumption that one can validly speak of God's purposes—or of God—is not something that can be blithely accepted as a "given" to which a theory of teleology must adjust. These issues require philosophic argumentation and objective validation by reference to an objective standard.

Pragmatism and analytic philosophy see philosophy as a melange of independent, loosely related microtopics, any one of which can be singled out for consideration while putting all the others on hold. Objectivism holds that philosophy, like knowledge generally, has a hierarchical structure such that the discussion of derivative topics, like teleology, presupposes the settled establishment of the entire structure beneath them—which in this case means all the essentials of metaphysics and epistemology.

Wright's pragmatism leads him to state:

> It is clear that the demonstration of the goal-directedness of something's behavior does not involve us at all in a discussion of the internal structure of that something . . . [It] could be an organism, a mechanism, a lump of quartz or a forest fire.[27]

Wright here opposes the essence of my entire book, as stated in its title: the biological basis of teleological concepts. Wright's position is that "natural selection. . . is the sort of principle that underwrites consequence-etiological explanations." My position is that the conditional nature of life gives rise to natural selection and that accordingly natural selection is the principle—the only principle—that underwrites consequence-etiology. Further, I hold that consequence-etiology (goal-causation) is meaningful only in the context of self-generation and value-significance. Wright actively opposes making value-significance part of the analysis, as is clear from his criticism of Ayala's use of the term "utility" in the quoted passage. He also would oppose bringing in self-generation, since he regards the motion of electrons as being teleological.

I have been concerned here to distinguish my own view from Wright's, and to defend my view on the points where we differ. But in so doing, I run the risk of conveying the misimpression that I find no value in Wright's work. On the contrary, he is one of the very few writers on the topic who actually grasps the general facts of the case. That he has seen only part of the truth, in my view, is due to his being the perhaps unwitting captive of Kantian epistemology. But in comparison to such writers as Pap, Nagel, Braithwaite, Wiener, and a host of lesser contemporaries, Wright towers like a giant. One need only read Wright's contemporary critics[28] to appreciate how reality-centered he is by comparison. The critics simply do not seem to understand the obvious and incontrovertible facts that Wright has identified.

Wright's book has been widely read, discussed, and criticized, but I have found only one of the criticisms to be cogent. That criticism is raised by Christopher Boorse (*Philosophical Review*, January 1976) in an otherwise hopeless paper defending Nagel-type views against Wright. The criticism Boorse raises depends on a counterexample. Although the method of attack by counterexample is grotesquely overdone in contemporary analytic philosophy, and I have grave doubts about the general validity of arguing from counterexamples, there is a certain point to this one, because it highlights the problems stemming from Wright's opposition to bringing value-significance into his analysis. Boorse writes:

> By failing to make any requirement that functions benefit their bearers, the analysis also creates unwanted func-

tions of the following sort. Obesity in a man of meager
motivation can prevent him from exercising. Although
failure to exercise is a result of the obesity, and the obesity continues because of this result, it is unlikely that prevention of exercise is its function.[29]

This is not a fatal objection to Wright's view, since (as
Boorse himself immediately notes) Wright may reply that
there has been no selection for either sedentariness or obesity. In fact, the case appears to be a counterexample only
because we have arbitrarily abstracted out a limited portion of
the whole sequence of causes and effects. The whole
sequence is: X's obesity leads to X's sedentary mode which
leads to X's greater obesity, which impairs X's health, which
leads to selection *against* the whole obesity-sedentary syn-
drome. Still, the example shows that Wright's omission of
value-significance tends to make us think of all cyclical processes—good and bad alike—as prima facie teleological. My
own analysis certainly makes it much clearer that such vicious
cycles are nonteleological, since they lack survival value or are
actually detrimental to the organism.

ANDREW WOODFIELD

I turn now to Andrew Woodfield's book, *Teleology*, which
appeared in 1976.

Like most authors on the subject, Woodfield makes a basic
distinction between functions and goals—a position I have
already criticized. And, as with Wright, Woodfield's position
on functions is much closer to my own than is his position on
goals. "Functional explanations, even of artifacts, do not say
merely 'This does some good'; they say 'This is here *because* it
does some good'."[30]

Woodfield seems to come closer to my position than
Wright does in one respect: Woodfield sees value-significance
as essential to the analysis of teleology. But, as we shall see, his
concept of "good" is much wider than mine, with the result
that this similarity in our positions tends to evaporate.

In regard to goal-causation (or consequence-etiology),

Woodfield attempts to deal with the problem of backward-causation by retreating from particular instances to general connections:

> The functional TD [teleological description] is not referring to any particular heart, but to hearts in general, or to the heart, *qua type* of organ. Consequently, it is not about any particular occasion of beating, but about heart-beating in general. It asserts "Hearts in general beat because beating (characteristically) contributes to blood circulation, and blood circulation is good for the organism."[31]

While we can indeed take this "tenseless" perspective, I have argued that we can also talk about "any particular occasion of beating" and observe that its occurrence has been made possible causally by previous occurrences. Woodfield seems to agree, at least in part: "Functional TD's sum up a number of historical facts by fudging time-references, thereby creating the illusion that the cause of the present beating is the fact that it will have a beneficial effect."[32]

Woodfield also makes the point that there are two levels of selection here: ontogenetic and phylogenetic:

> The question arises, which organisms are benefited by which hearts? This vague generalization quantifies loosely over not only hearts and heartbeats, but also over the owners of the hearts. There are two ways of unpacking it [i.e., phylogenetically and ontogenetically]. Nearly every heart that ever beats is present in its owner's body because other hearts, present in other bodies, have beaten in the past. By so doing, they have contributed to the survival of their owners, who were the ancestors of this organism. If it hadn't been for the ancestral hearts, this heart would never have come into existence. Thus it would be true of any heart (except a mythical "first-generation heart") that it beats because heart-beating has contributed to the preservation of the species of which this heart-owner is a member. . . .This is the "phylogenetic" interpretation. But the TD can be given an "ontogenetic" interpretation where "S" stands for an individual. . . .

> At any time *t* in the life of an individual with a heart. . .it
> is true to say that his heart has contributed to his surviv-
> ing up until *t* by circulating his blood. If the heart had
> not beaten prior to *t*, the owner would not be alive at *t*.
> Therefore, every heart beats because its own earlier beat-
> ing has helped its owner not to die. . . .The explanans is
> "Because beating contributes to circulation, which is
> good"—a tenseless general statement about event-types....
>   Phylogenetic and ontogenetic interpretations can be
> given for behavior-functions too.[33]

Thus this section of Woodfield's presentation is in accord
with the theory I have presented, and there are other similar
statements in his book which I would endorse (although not
in Woodfield's context).

In regard to goal-directed action, Woodfield holds that to
use the term "goal" is to imply that what is being attained is
in some sense good and that the organism has an "internal"
(i.e., conscious) representation of the goal. He attacks what
he calls "externalist" theories—i.e., theories which hold that
consciousness is not an essential element in goal-directed
action. "According to the externalist, non-conscious goals are
logically prior to conscious ones."[34] My theory thus qualifies
as "externalist."

Woodfield's overall view is that purposeful and vegetative
actions are to be analyzed exactly as I have done, but that we
should not integrate the two types of causation into one concept—"goal-directed action." His grounds are that the
involvement of consciousness is a fundamental differentiating
feature. "Goals . . . are mental entities, living permanently
inside intentional brackets."[35]

One might wonder whether the difference between our
views is merely terminological. Does Woodfield merely mean
to use the word "goal" for those things that I denote by the
word "purpose"—i.e., for specifically conscious ends? It is difficult to give an unequivocal answer, because (a) Woodfield
does not define his terms, (b) he does not clearly separate
cases of vegetative and conscious action, and (c) he does not
grasp the difference between the perceptual and conceptual

levels of consciousness. "In crediting the rat with perceptions," he writes, "one is already crediting it with beliefs. In perceiving that the food has moved to the left, the rat acquires the belief that the food has moved to the left."[36] He even asserts that "The animal may have false beliefs."[37] Accordingly, Woodfield analyzes what he calls "goals" not just in terms of consciousness, but also in terms of beliefs: "The presence or absence of the psychological verb 'believes' marks a crucial dividing line."[38]

I agree that "beliefs" do mark a dividing line, but clearly that line is the one that separates man's conceptual faculty from the purely perceptual-level faculties of subhuman animals. One cannot subordinate the distinction between the conscious and the non-conscious to that between the conceptually conscious and the perceptually conscious. "Beliefs" are not even fundamental to the conceptual level. Beliefs are hypotheses; they arise only after a conceptual consciousness has reached the fairly advanced stage of recognizing such things as error, evidence, and doubt. There is a distinction between beliefs and conclusions or thoughts, simpliciter. But even thoughts or conclusions are derivative phenomena; they express propositions. The basic form of a proposition is: "S is P," where "S" and "P" denote *concepts*. Rats, not possessing concepts, cannot form propositions, cannot think, cannot err, cannot distinguish the true and the false, cannot have doubts, cannot weigh evidence, and therefore cannot have beliefs. And if Woodfield is willing to endow rats with true and false beliefs, and if he considers this as assumed in describing a rat's behavior as goal-directed, one wonders what he would say about lower animals, such as frogs, worms, oysters, and sponges—and about plants and mitochondria.

Woodfield, like Wright, is radically undercut by the analytic/synthetic dichotomy, along with its metaphysical corollary: the dichotomy between "necessary" and "contingent" facts. Consider the following passage:

> Richard Sorabji[39] has invented a hypothetical example to show that *logically there could be* functions which had no

survival or reproductive value. He imagined an animal having a mechanism inside it which comes into operation only when some lethal damage has occurred, like a major heart-attack. The mechanism has the effect of shutting off pain from the damaged area. Unlike the ordinary mechanism of sensory adaptation, it does not promote the survival either of the individual or the species. But it would surely be correct to say that shutting off pain was the function of the mechanism. Therefore, survival and reproduction are not the only possible ends (p. 117, my emphasis).[40]

Observe the methodology here. A state of affairs counter to the facts of reality is dreamed up, then on no basis other than the word "surely," one announces how it is to be classified; then one uses that, entirely arbitrary, dictum to conclude that analyses based on the facts as they actually are have been refuted. What could Sorabji say in response to my, extensively argued, claim that "surely it would *not* be correct to say that shutting off pain was the function of the mechanism"? Presumably, he could only reply that his "intuitions" do not match mine. Such is the subjectivism engendered by contemporary methodology.

In fact, philosophy, as opposed to religion, cannot deal in anyone's "intuitions." Such "intuitions" are merely the feelings produced by subconscious, automatized ways of conceptualizing phenomena—conceptualizations that may have been made incorrectly. A person brought up on the "stripes" vs. "non-stripes" system of biological classification might find his "intuitions" rebelling against the proposal to divide organisms between plants and animals. *Proper* conceptual classifications are determined by reference to the logical hierarchy of concepts—which requires being able to reduce higher abstractions, such as "function" and "goal," to their base in perceptual reality. In the present instance, this means showing what facts of reality give rise to our need of the concept "goal." Those facts are: that organisms are capable of self-generated action, that they face the alternative of their existence or non-existence, and that their continued existence depends on their successful performance of specific courses of action. Continuing the passage from Woodfield:

> There are other ends over and above mere life, such as health, pleasure and the absence of pain. Sorabji then cites some of Aristotle's examples of "luxury functions," i.e., functions which are not essential for survival, but which do the organism good in some way.[41]

This is put tendentiously. Rather than describing health, pleasure, and the absence of pain as "ends over and above mere life," one should say that these ends—insofar as they are ends—are *subsumed by* life. In the body of this book I have shown in considerable detail how the pleasure-pain mechanism of animals is a survival instrument because it serves to guide the animal toward that which furthers its survival and away from that which threatens it. Precisely because Sorabji's hypothetical pain shut-off mechanism is irrelevant to survival, it is irrelevant to teleology. The fact that Woodfield includes *health* in his list of items that are irrelevant to "mere life" only shows the question-begging nature of the "mere." The concept "health" cannot be grasped or defined except in relation to the requirements of life.

Woodfield himself does not entirely agree with Sorabji's position. He is here presenting Sorabji's counterexample mainly as a "problem" for discussion. Later he states that "life is a privileged member of the class of biological ends."[42] But he immediately adds: "The others [reproduction, pleasure, pain-elimination] do not causally contribute to it [life]"[43]—as if an animal could live without having being born, or could survive bereft of the guidance of pleasure and pain. Woodfield ultimately concludes, correctly, that Sorabji's pain shut-off mechanism is not actually teleological ("It would be false to say of Sorabji's mechanism that it exists *because* it shuts off pain"[44]). But the analytic/synthetic dichotomy prevents him from building this fact into his analysis of teleological concepts.

The reality-negating effect of the analytic/synthetic dichotomy is evident in Woodfield's next statements:

> It is probable that no examples of luxury functions actu-
> ally *exist* in nature, since natural selection does not favor
> features that confer no adaptive advantage. But this is
> merely a contingent matter. So long as it makes sense to
> think that animals could have good done to them in
> other ways, it is conceivable that they might have had
> parts which did them good in these other ways. For one
> who wishes to understand the meaning, the conceptual
> boundaries, of the term "function," it would be a mistake
> to consider only those features that happen to exist in
> nature. . . . To think that evolutionary theory can expli-
> cate the *meaning* of "function" is to get things back to
> front. The word "function" was in common use long
> before Darwin. . . . Sorabji's hypothetical case is impor-
> tant, because it shows that the sense of the word "func-
> tion" is independent of Darwin's great discovery that
> many or all natural functions yield to systematization
> within evolutionary biology. Indeed, insistence upon this
> point is essential in order to defend Darwin against the
> criticism that his "natural selection" hypothesis was a
> mere tautology.[45]

To engage in a colossal understatement: I disagree. Read-
ers familiar with Objectivist epistemology will understand the
grounds of that disagreement. For those unfamiliar with the
Objectivist theory, I strongly recommend "The Analytic-Synthetic Dichotomy" by Leonard Peikoff (included in *Introduction to Objectivist Epistemology*). The essential points Peikoff
makes are: (1) the meaning of a concept includes all the characteristics of the concept's referents, not just those characteristics stated in the concept's definition, and (2) the concept of
"contingent" facts—facts which "happen to exist in nature"
but which allegedly are not "necessary"—is an invalid rem-
nant of religious metaphysics.

If Woodfield makes damaging concessions to the analyt-
ic/synthetic dichotomy in his treatment of "functions" he
capitulates entirely to the alleged supremacy of the "logically
possible" over "what happens to be the case" in his treatment
of "goals":

There is no contradiction in the idea that animals might instinctively aim at things which are bad for them, or neither good nor bad. Whenever an animal strives after some G which is not a basic goal [i.e., life] or a means to a basic goal, the externalist is committed to saying that G is a derived goal or a displacement goal. He thus commits himself to a factual claim about the animal's past history, that S acquired the goal G because G used to be associated with a basic goal, or because of a conflict between basic goals. Many acquired goals do, in fact, arise in these ways. But it is not conceptually necessary that all non-basic goals should be derived from a set of basic goals. . . . It is not impossible for an animal to have the goal of moving its shadow from A to B, or of going to the northwest corner of the room for its own sake, even though these goals are not, and never have been, beneficial to the organism or associated with anything beneficial.[46]

Thus, the result of the analytic/synthetic dichotomy in Woodfield's work is the same separation of "goal" and "life" that we saw in Wright's. My own position is "externalist" in the sense that I regard all teleological concepts—"goal" in particular—as rooted in the biological alternative of life vs. death, not in psychological alternatives such as pleasure vs. pain or desire vs. aversion. Woodfield's entire attack on "externalism" (pp. 154-159) consists in the double error of wielding the analytic/synthetic dichotomy against a straw man: the "plasticity" approach of Braithwaite, Nagel, Sommerhof, et al. Although he does recognize that goal-directed action implies that the action occurs because its effects are good for the organism, he would certainly oppose my central thesis, which is summed up in this passage from Ayn Rand:

Only a *living* entity can have goals or can originate them. And it is only a living organism that has the capacity for self-generated, goal-directed action. On the *physical* level, the functions of all living organisms, from the simplest to the most complex—from the nutritive function in the single cell of an amoeba to the blood circulation in the body of a man—are actions generated by the organism

235

itself and directed to a single goal: the maintenance of the organism's *life*. . . . Metaphysically, *life* is the only phenomenon that is an end in itself: a value gained and kept by a constant process of action. Epistemologically, the concept of "value" is genetically dependent upon and derived from the antecedent concept of "life." To speak of "value" as apart from "life" is worse than a contradiction in terms. "It is only the concept of 'Life' that makes the concept of 'Value' possible."[47]

NOTES

CHAPTER I

1. See H. and H. A. Frankfort, *Before Philosophy* (Baltimore, 1949), p. 24.
2. Aristotle reports Thales to have said also that everything is full of gods, but Thales may not have meant this in the animistic sense: "It seems likely, therefore, that in saying things are full of gods, Thales was not making a theological statement. Paradoxical as it may sound, he was tacitly denying divine causality. He meant that things, in order to be moved and to change, do not require force applied to them from the outside by the gods but move of themselves, by a natural [non-conscious?] force within them." (W.T. Jones, *A History of Western Philosophy* (New York, 1952), p. 34.
3. Hans Jonas, *The Phenomenon of Life* (New York, 1966), pp. 7-8.
4. "Consciousness" is an axiomatic concept; it can be defined only ostensively. See Ayn Rand, *Introduction to Objectivist Epistemology* (New York, 1979), chapter 6.
5. Niko Tinbergen, *Social Behavior in Animals* (London, 1964), pp. 8-14.
6. "Imprinting" seems to be such a borderline case: the "programming" for imprinting appears to be innate, and once acquired, the imprinting is relatively inflexible, but the particular object on which the animal imprints is learned in its early sensory experiences.
7. I leave aside the issue of *sub*conscious purposes in the psychologist's sense.

CHAPTER II

1. Ranier Schubert-Soldern, *Mechanism and Vitalism*, trans. and ed. by Philip G. Fothergill (Notre Dame, 1962), p. 57 and p. 219. See also Allan Gotthelf's argument that Aristotle's "final cause" involves an "irreducible potential for form" (*Review of Metaphysics*, December 1976, pp. 226-254).
2. R. H. Haynes and P. C. Hanawalt, "Introductory Essay," *The Molecular Basis of Life* (San Francisco, 1968), p. 4. I would specify "complex *physical* system," since the formula does not apply to consciousness itself (i.e., one cannot consider consciousness to be a "system" in this sense).
3. George G. Simpson, *This View of Life* (New York, 1964) p. 105.
4. Edmund W. Sinnott, *Cell and Psyche* (New York, 1961), pp. 49-50.
5. *Ibid.*, p. 59.
6. The last property is redundant, but including it makes the example simpler.

CHAPTER III

1. Magda Arnold, *Emotion and Personality* (2 Vols.: New York, 1960), I, 173.
2. A more adequate explanation of the role of volition in purposeful action would require a presentation of the Objectivist theory of free will. See Ayn Rand, *The Virtue of Selfishness* (New York, 1964), pp. 19-22.
3. R. B. Braithwaite, *Scientific Explanation* (Cambridge, 1960), pp. 324-325. See also Ernest Nagel, *The Structure of Science* (New York, 1961), p. 25.

CHAPTER IV

1. George Wald, "The Origin of Life," *Scientific American*, CXCI, No. 2 (1954), pp. 49-50.
2. William T. Keeton, *Elements of Biological Science* (New York, 1969), p. 89.

3. Albert Lehninger, "How Cells Transform Energy," *Scientific American*, CCV, No. 3 (1961), p. 71.
4. William A. Pryor, "Free Radicals in Biological Systems," *Scientific American* (August, 1970), p. 73.
5. Keeton, *Elements of Biological Science*, p. 180.
6. Carl J. Wiggers, "The Heart," *Scientific American*, CXCVI, No. 5 (1957), p. 75.
7. I owe this concept to Ayn Rand (*Atlas Shrugged* [New York, 1957], p. 1013), see passage cited *infra*, p. 62-63.
8. Plato, *Laws*, trans. by A. E. Taylor, in E. Hamilton and H. Cairns, Plato: *Collected Dialogues* (New York, 1961), p. 1451, 895c.
9. Aristotle, *De Anima*, Book II, I, 412b18, trans. by J. A. Smith, in *The Basic Works of Aristotle*, ed. by Richard McKeon (New York, 1941). I have translated "kinesis" as "action" rather than "movement."
10. *Physics*, Book VIII, 4, 255a5-8 (trans. by R. P. Hardie and R. K. Gaye).
11. See also V. R. Young and N. S. Scrimshaw, "The Physiology of Starvation," *Scientific American*, CCXXV, No. 4 (1971), pp. 14-21.
12. See Ernst Mayr's discussion of "programmed" biological action in "Cause and Effect in Biology," *Science*, CXXIV, (November, 1961), pp. 1503-1504.
13. R. O. Kapp, "Living and Lifeless Machines," *British Journal for the Philosophy of Science*, V (August, 1954), p. 101.
14. Harold F. Blum, *Time's Arrow and Evolution* (Princeton, 1968), chap. iii, *passim*.

CHAPTER V

1. Braithwaite, *Scientific Explanation*, p. 329.
2. Israel Scheffler, *The Anatomy of Inquiry* (Indianapolis, 1963), chapter 9.
3. For similar views, see Michael A. Simon, *The Matter of Life* (New Haven, 1971), p. 81, and John Canfield, "Teleological Explanations in Biology," *British Journal for the Philosophy of Science*, XIV (1964), pp. 290-291.

4. Ayn Rand, *The Virtue of Selfishness* (New York, 1964), p. 15.

5. The following application of Miss Rand's statement is my own, and she is not responsible for any errors it may contain.

6. Simpson, *This View of Life*, p. 173.

7. Rand, *Atlas Shrugged*, pp. 1012-1013.

8. Lehninger, "How Cells Transform Energy," p. 62.

9. It has been claimed that a counterexample to this principle is presented by the case of organisms which during some part of their life-cycle can seemingly exist without performing self-sustaining actions. For example, there are viruses that can exist in a crystallized state which precludes the possibility of any internal action. But if life is a process of self-generated, self-sustaining action, viruses do not qualify as living organisms. It is true that such a virus as the tobacco mosaic virus contains a nucleic acid core which can serve as a template for the reproduction of other such viruses by a host cell, but this no more implies that the virus is alive than the fact that a mimeograph stencil can be used to make copies of itself implies that the stencil is alive. Other apparent counterexamples are generally cases in which a true living organism goes through a dormant, inactive state (e.g., bacterial endospores); however, it appears that in such cases internal action does not cease totally—rather, it continues at a highly reduced rate. Secondly, it is clear that such dormant states are dependent upon, and can be understood only in the context of, a complete life-cycle necessarily containing active phases. Finally, no counterexamples could ever invalidate the general principle that life is conditional upon action.

10. Walter Bock, Columbia University Department of Zoology, personal communication. (Dr. Bock is not responsible for the exact formulation given here of these three features.)

11. Rand, *The Virtue of Selfishness*, p. 17.

12. Th. Dobzhansky, *The Biology of Ultimate Concern* (New York, 1967), p. 21.

13. Th. Dobzhansky, *Genetics of the Evolutionary Process* (New York, 1970), p. 6.
14. G. G. Simpson, *The Major Features of Evolution* (New York, 1953), pp. 160-161. Actually, "adaptation" should be defined in terms of those benefits to survival that have *caused* the selection of the feature in question. See Robert Brandon, "Biological Teleology: Questions and Explanations," *Studies in the History and Philosophy of Science,* vol. 12, no. 2 (1981), p. 97ff.
15. Braithwaite, *Scientific Explanation,* p. 329.
16. A. Rosenblueth, N. Wiener, J. Bigelow, "Behavior, Purpose, and Teleology," *Philosophy of Science,* X (1943), pp. 18-24.
17. Richard Taylor, "Purposeful and Non-purposeful Behavior: A Rejoinder," *Philosophy of Science,* XVII (1950), pp. 327-332.
18. Perhaps we could not say that the rock's goal is achieved by goal-directed action, since the requisite feedback control does not exist, but the final condition would be a *goal* nonetheless.
19. The translation of *to hou heneka* as "the for what" is given by John H. Randall, Jr., *Aristotle* (New York, 1960), p. 124; "that for the sake of which" is more standard—see, e.g., W. D. Ross's translation in *Metaphysics,* V, 2, 1013a30-31.

CHAPTER VI

1. George C. Williams, *Adaptation and Natural Selection* (Princeton, 1966), pp. 12-13.
2. Francisco J. Ayala, "Teleological Explanations in Evolutionary Biology," *Philosophy of Science,* XXXVII (March, 1970), p. 12.
3. Braithwaite, *Scientific Explanation,* pp. 324-325.
4. In some cases the goal may more nearly approximate the goal-object. This occurs when the goal is the creation of some entity—for instance, the creation of a sculpture by an artist. Technically, it is perhaps still true that the goal is not just the existence of the sculpture, but also its con-

templation by the artist and other viewers.

5. —and, in the case of man, by introspective awareness of the actions of his consciousness, which in turn represent (or depend, ultimately, on) prior awareness of external objects.

6. I use "extrapolation" throughout as a wide, general term to include many, somewhat disparate, mental processes—from simple perceptual association through elaborate conceptual processes of induction and deduction. All these processes, however, involve a projection of the future based on the past.

7. Scheffler, *The Anatomy of Inquiry*, pp. 115-116.

8. This use of the distinction between proximate and ultimate cause was suggested by John Herman Randall, Jr., Columbia University Department of Philosophy. A similar distinction has been suggested by Mayr, ("Cause and Effect in Biology," p. 1503).

9. William Dray, *Laws and Explanation in History* (Oxford, 1957), p. 72.

10. Albert Lehninger, "Energy Transformation in the Cell," in *Readings in the Life Sciences* (*Scientific American Resource Library*), II (San Francisco, 1970), p. 558.

11. Simpson, *This View of Life*, p. 113. See also Colin S. Pittendrigh, "Adaptation, Natural Selection, and Behavior" in *Behavior and Evolution*, ed. by A. Roe and G. G. Simpson (New Haven, 1958), p. 395.

12. Dobzhansky, *Genetics of the Evolutionary Process*, pp. 5-6.

CHAPTER VII

1. Ayala, "Teleological Explanations in Evolutionary Biology," p. 2.

2. In fact, this struggle is not only directed against other, competing organisms, but also, and primarily, it is the individual struggle to obtain the means of survival from the natural environment.

3. Wald, "The Origin of Life," pp. 49-50.

4. Ayala, "Teleological Explanations in Evolutionary Biol-

ogy," p. 9.

5. *Ibid.*, p. 10.

6. I do believe, however, that Ayala's analysis does tacitly assume the kind of goal-causation that is being defended in this book.

7. Dobzhansky, *The Biology of Ultimate Concern*, pp. 24-25.

8. Pittendrigh, "Adaptation, Natural Selection, and Behavior," pp. 396-397.

9. Simpson, *This View of Life*, pp. 104-105 and pp. 118-119.

10. Williams, *Adaptation and Natural Selection*, p. 31.

11. Keeton, *Elements of Biological Science*, p. 81.

12. In addition ATP is used to drive certain energy-requiring steps in the respiratory process itself. (Of course, the "justification" of this is that more molecules of ATP are produced in respiration than are consumed, for a resulting net gain of ATP.)

13. Possibly, phototropism requires no extra energy at all: if phototropism affects only the *direction* of growth, not its overall rate, then the only extra energy required is that used in the ontogenetic development of the auxin mechanism, including auxin synthesis, plus any energy that might be required to transport the auxins to the shaded side of the stem. See chapter 4.

14. Charles Darwin, *The Origin of Species*, Mentor Edition (New York, 1958), p. 29.

15. R. A. Fisher, *The Genetical Theory of Natural Selection* (New York, 1958), pp. 19-20.

16. Arthur Pap, *An Introduction to the Philosophy of Science* (New York, 1962), p. 361.

17. Carl J. Wiggers, "The Heart" in *Readings in the Life Sciences* (*Scientific American Resource Library*), II (San Francisco, 1970), p. 491.

18. Walter J. Bock and Gerd von Wahlert, "Adaptation and the Form-Function Complex," *Evolution*, 19 (1965), p. 286.

19. Ayala, "Teleological Explanations in Evolutionary Biology," p. 12.

20. Dobzhansky, *Genetics of the Evolutionary Process*, p. 95.

21. *Ibid.*, p. 2. Note that as Walter Bock and G. von Wahlert have pointed out ("Adaptation and the Form-Function Complex," *Evolution*, pp. 290-291), such definitions omit reference to the *cause* of the differential reproduction of alternative genetic units, and they have offered a more exact definition which, unfortunately, is too technical to be introduced here. But see Robert Brandon, "Adaptation and Evolutionary Theory," *Studies in the History and Philosophy of Science*, 1978, reprinted in Eliot Sober (ed.), *Conceptual Issues in Evolutionary Biology* and "Biological Teleology: Questions and Explanations," 1981, pp. 9ᵀ 105.

22. Ayala, "Teleological Explanations in Evolutionary Bioı ogy," p. 4.

23. Walter Bock and G. von Wahlert, "Two Evolutionary Theories—A Discussion," *British Journal for the Philosophy of Science*, XIV, No. 54 (1963), p. 143.

24. Walter Bock, "Comparative Morphology in Systematics," *Systematic Biology* (Proceedings of an International Conference, National Academy of Sciences, Washington, D. C., 1960), p. 440.

25. J. B. S. Haldane, *The Causes of Evolution* (Ithaca, N. Y., 1966), pp. 142-143.

26. Gavin de Beer, *Evolution* (London, 1964), p. 22.

27. Nagel, *The Structure of Science*, p. 402. (Nagel, of course, uses this phrase to characterize the view he too opposes.)

CHAPTER VIII

1. And, in the case of man, the extrapolations are based on forms of value-significance more complex than pleasure or pain—i.e., abstractly held goals, values, and moral principles.

2. This point was suggested to me by George Reisman in personal conversation. Also, a quite similar point is made by Ayn Rand: "In what manner does a human being discover the concept of 'value'? By what means does he first become aware of the issue of '*good or evil*' in its simplest

form? By means of the physical sensations of *pleasure* or *pain.* Just as sensations are the first step of the development of a human consciousness in the realm of *cognition* so they are its first step in the realm of *evaluation.*" ("The Objectivist Ethics," *The Virtue of Selfishness*, p. 17.)

3. *The American Heritage Dictionary*, 1969, 1970.
4. *De Anima*, II, 3. 414b5-6.
5. Arnold, *Emotion and Personality*, I, p. 238.
6. Fisher, *The Genetical Theory of Natural Selection*, p. 151.
7. William James, *Principles of Psychology* (New York, 1923), I, pp. 143-144.
8. Curt P. Richter, et al., "Nutritional Requirements for Normal Growth and Reproduction in Rats," *American Journal of Physiology*, CXXII (1938), pp. 733-744.
9. James, *Principles of Psychology*, p. 144.
10. Rand, *The Virtue of Selfishness*, pp. 17-18 and p. 27.
11. Jonas, *The Phenomenon of Life*, p. 126. (I do not endorse Jonas' implication that "urges"—i.e., desires—are present on the vegetative level.)
12. In fact, as argued in chapter 4 (*supra*), the concepts denoting value-significance (e.g., "benefit" and "harm") are not applicable to inanimate objects.
13. *Physics*, II, 8, 198b18-20.
14. E. E. Stone, "Rain," *The World Book Encyclopedia* (1959), p. 6785.
15. Lehninger, "Energy Transformation in the Cell," p. 558.
16. *Ibid.*
17. Also, as argued earlier, the clock's behavior is not self-generated. On epistemological grounds, it would probably be necessary to withhold the term "goal-directed" from the behavior of any such device that did not actually count as a living organism—but this is a different matter. The "epistemological grounds" would be that the concept of "goal-directedness" should not be taken out of the biological context that gave rise to it. Contrary to contemporary notions, the mere fact that something satisfies a concept's definition does not automatically and invariably make it a proper referent of the concept. The

function of a definition is not to serve as the absolute cri-
✦ terion of class-inclusion. See Appendix, and Rand, *Introduction to Objectivist Epistemology*, chapter 5.

18. Frederick Warburton, Columbia University Department
of Biology, personal communication.

19. A wider context for the reduction of qualitative differences to quantitative ones is provided in Rand's theory
of similarity as reducible to differences of measurement.
See *Introduction to Objectivist Epistemology*, ch. 1 & 2.

20. T. A. Goudge, *The Ascent of Life* (Toronto, 1961), p. 193.

21. Ayala, "Teleological Explanations in Evolutionary Biology," II.

CHAPTER IX

1. Dobzhansky, *Genetics of the Evolutionary Process*, p. 97.

2. We cannot infer with *certainty* that merely because the
behavior exists, it must be adaptive to the particular end
of protecting the young, or that it is adaptive at all. We
can however infer that there is an extremely high probability that the behavior is adaptive, and adaptive for this
end.

3. I mean no physical mechanism analogous to the DNA
mechanism which coordinates the actions of the cell.

4. J. B. S. Haldane, "Population Genetics," *New Biology*,
XVIII (1955), pp. 34-51.

5. Williams, *Adaptation and Natural Selection*, p. 252.

6. *Ibid.*, p. 160.

7. Ernst Mayr, *Populations, Species, and Evolution* (Cambridge, MA, 1970), p. 115.

8. *Ibid.*, p. 387.

9. Simpson, *The Major Features of Evolution*, p. 166.

10. Konrad Lorenz, "Beitrge zur Ethologie sozalier Corvi-
den," *Journal fr Orinthologie*, LXXIX (1931), pp. 67-120;
"Der Kumpan in der Umwelt des Vogels," *Journal fr
Orinthologie*, LXXXII (1935), pp. 137-213.

11. Niko Tinbergen, "Foot-Paddling in Gulls," *British Birds*,
LV (1962), pp. 117-120.

12. A. D. Blest, "The Function of Eyespot Patterns in the Lepidoptera," *Behaviour,* XI (1957), pp. 209-256.
13. E. Cullen, "Adaptations in the Kittiwake to Cliff Nesting," *The Ibis,* XCIX (1957), pp. 275-302.
14. Niko Tinbergen, "Behavior and Natural Selection," in *Ideas in Evolution and Behavior,* ed. by John A. Moore, Proceedings XVI International Congress of Zoology, vol. VI (Garden City, N. Y.), p. 528 and p. 533.
15. And to the extent this "play" behavior has been artificially selected by man in breeding playful cats, it may be viewed as an adaptation to the human environment.
16. Judith Berliner, Yale University Department of Biology, personal communication.
17. N. E. Miller, "Theory and Experiment Relating Psychoanalytic Displacement to Stimulus-Response Generalization," in *The Study of Personality,* ed. by H. Brand (New York, 1954).
18. D. O. Hebb, *A Textbook of Psychology* (Philadelphia, 1958), pp. 235-238.

CHAPTER X

1. Aristotle, *Physics,* II, 198b29, trans. by R.P. Hardie and R.K. Gaye, in *The Basic Works of Aristotle,* ed. by Richard McKeon (New York, 1941).
2. Simpson, *This View of Life,* pp. 51-52.
3. Darwin, *The Origin of Species,* pp. 87-88.
4. We cannot, however, include the development of bacterially induced diseases, since these are not actions of the diseased organism, but of the invading bacteria. The defensive actions taken by the organism (e.g., antibody production) are goal-directed toward its survival. Tumor growth, at least on one prevalent theory, is a malfunction of the cell's own DNA programming.
5. For a full refutation of the base of this sort of objection to my thesis, see "The Analytic-Synthetic Dichotomy," by Leonard Peikoff in Rand's *Introduction to Objectivist Epistemology* (New York, 1979).

6. G. C. Field, "The Place of Definition in Ethics," reprinted in *Readings in Ethical Theory*, ed. by Wilfrid Sellars and John Hospers (New York, 1952), pp. 92-102.

7. W. K. Frankena, "The Naturalistic Fallacy," reprinted in *Readings in Ethical Theory*, pp. 103-114.

8. *Ibid.*, p. 110.

9. Some concepts may be definable both by ostensive means and by a formal statement of genus and differentia. E.g., "square" might be definable by giving examples and also as "a rectangle with four equal sides." Since such concepts would still be eliminable in favor of their non-ostensive definitions, they would have to be done away with on the eliminativist premise.

10. Nagel, *The Structure of Science*, p. 366.

11. Rand, *Introduction to Objectivist Epistemology*, p. 59.

12. *Ibid.*, pp. 94-95. For further discussion of the principle of fundamentality and standards for conceptual classification, see Rand, *Introduction to Objectivist Epistemology* (New York, 1979), especially chapters 5 and 7.

CHAPTER XI

1. Rodolfo Margaria, "The Sources of Muscular Energy," *Scientific American*, CCXXVI, No. 4 (1972), p. 84.

2. Hugh Davson, *A Textbook of General Physiology* (Boston, 1959), p. 154.

3. Goudge, *The Ascent of Life*, p. 193.

4. N. W. Pirie, "Concepts out of Context," *British Journal for the Philosophy of Science*, II (1951), p. 278.

5. Keeton, *Elements of Biological Science*, p. 258.

6. *Ibid.*, p. 95.

7. *Ibid.*, p. 103.

8. *Ibid.*, p. 89.

9. Th. Dobzhansky, *Genetics of the Evolutionary Process*, p. 4.

10. A more "efficient" means is one costing the organism less energy to attain the same end. See Bock and von Wahlert, "Adaptation and the Form-Function Complex," pp. 286-290, for a similar point.

11. Basic to such a translation would be a statement connecting the term "life" or "survival" with mechanical terms.
12. See W. D. Hamilton, "The Genetical Evolution of Social Behavior," *Journal of Theoretical Biology*, 1964.
13. Herbert H. Ross, *Understanding Evolution* (Prentice-Hall, Englewood Cliffs, N.J., 1966), p. 106.
14. Mayr, *Populations, Species, and Evolution*, pp. 147-148.
15. B. Wallace, "Genetic Divergences of Isolated Populations of *Drosophila melanogaster*," Proceedings of the 9th International Conference on Genetics, *Caryologia*, VI (Suppl., 1954), pp. 761-764.
16. V. G. Dethier, and E. Stellar, *Animal Behavior* (Englewood Cliffs, N. J., 1961), p. 75.
17. On the psychological (and, ultimately, biological) function of art, see Rand, "The Psycho-Epistemology of Art" in *The Romantic Manifesto*.
18. Theodosius Dobzhansky, *The Biological Basis of Human Freedom* (New York, 1956), pp. 86-87.
19. Theodosius Dobzhansky, "The Road Traversed and the Road Ahead," in *Readings in Biological Science*, ed. by I. W. Knobloch (New York, 1967), p. 458.
20. George G. Simpson, *The Meaning of Evolution* (New York, 1971), p. 259 and p. 258.
21. Desmond Morris, *The Naked Ape* (New York, 1967).
22. Simpson, *The Meaning of Evolution*, p. 261. Actually, the ability to use language is a derivative of the fundamental issue: man's ability to use reason.
23. Rand, *The Virtue of Selfishness*, p. 21.
24. Ayala, "Teleological Explanations in Evolutionary Biology," p. 11.
25. Simpson, *The Meaning of Evolution*, p. 283.
26. *Ibid.*

APPENDIX

1. Harry Binswanger, "Validating Ethics," *The IREC Review*, December 1966, pp. 2-3.
2. Francisco Ayala, "Teleological Explanations in Evolution-

ary Biology," *Philosophy of Science*, 37, 1970.

3. Larry Wright, "Explanation and Teleology," *Philosophy of Science*, 39, 1972.

4. Larry Wright, "Functions," *Philosophical Review*, 82, 1973.

5. Larry Wright, *Teleological Explanations* (Berkeley: 1976).

6. *Ibid.*, p. 94. The references are: Hempel, "The Logic of Functional Analyses," in *Symposium on Sociological Theory*, ed. L. Gross (New York, 1959); Canfield, "Teleological Explanations in Biology"; Sorabji, "Function," *Philosophical Quarterly* 14:289-302; Lehman, "Functional Explanations in Biology," *Philosophy of Science* 32:1-20; Gruner, "Teleological and Functional Explanations," *Mind* 75:516-526; Beckner, "Function and Teleology," *Journal of the History of Biology* 2:151-169. Wright also credits Beckner's book *The Biological Way of Thought* (Berkeley, 1968).

7. *Ibid.*, p. 39.

8. *Ibid.*, p. 56.

9. *Ibid.*, pp. 23-24.

10. *Ibid.*, p. 56.

11. *Ibid.*, p. 74.

12. *Ibid.*, p. 81.

13. *Ibid.*, p. 84.

14. *Ibid.*, p. 92.

15. *Ibid.*, p. 79.

16. *Ibid.*, p. 114.

17. *Ibid.*, p. 89.

18. *Ibid.*, pp. 41-42.

19. *Ibid.*, p. 46.

20. Aristotle, *Metaphysics* I, 1, 980a29, trans. by W.D. Ross, in McKeon.

21. Wright, p. 21.

22. *Ibid.*, p. 18.

23. *Ibid.*

24. *Ibid.*, pp. 18-19.

25. *Ibid.*, p. 70.

26. *Ibid.*, pp. 96-97.

27. *Ibid.*, p. 59.

28. Christopher Boorse, "Wright on Functions," *Philosophical*

Review, January 1976. Peter Achinstein, "Function Statements," *Philosophy of Science,* 44, 1977. Stephen Utz, "On Teleology and Organisms," *Philosophy of Science,* 44, 1977. Lowell Nissen, review of *Teleological Explanations, The Thomist,* April 1979. Lowell Nissen, "Wright on Teleological Descriptions of Goal-Directed Behavior," *Philosophy of Science,* 50, 1983. Douglas Ehring, "Teleology and Impossible Goals," *Philosophy and Phenomenological Research,* September 1986.

29. Boorse, p. 76.
30. Andrew Woodfield, *Teleology* (Cambridge: 1976), p. 136.
31. *Ibid.,* p. 137.
32. *Ibid.,* p. 209.
33. *Ibid.,* pp. 137-138.
34. *Ibid.,* p. 152.
35. *Ibid.,* p. 212.
36. *Ibid.,* p. 164.
37. *Ibid.,* p. 165.
38. *Ibid.,* p. 211.
39. Richard Sorabji, "Function," *Philosophical Quarterly,* 14, 1964.
40. Woodfield, p. 117.
41. *Ibid.*
42. *Ibid.,* p. 134.
43. *Ibid.*
44. *Ibid.,* p. 139.
45. *Ibid.,* pp. 117-118.
46. *Ibid.,* pp. 156-157.
47. Ayn Rand, *The Virtue of Selfishness* (New York: 1964), pp. 16-17.

INDEX OF AUTHORS